Naked Truth: Everything and Else

Sandeep Chavan

Published by SANDEEP CHAVAN, 2025.

NAKED TRUTH: EVERYTHING AND ELSE

First edition. July 22, 2025.

ISBN: 979-8231942954

Written by Sandeep Chavan.

Table of Contents

To the seekers who never settled for easy answers,and to the young minds across the globewho dare to question everything.

May you always remember —the truth was never hidden.It was simply waiting for alignment.

"The universe never spoke in words.It rippled.And those who listened with stillness,understood everything — and else."

Sandeep Chavan

Disclaimer

This book is not intended to diagnose, treat, or prescribe anything — materially, medically, spiritually, or scientifically — in the conventional sense.

Naked Truth presents a comprehensive reframing of reality through the lens of Universal Energy Dynamics (UED), a framework developed independently and shared here for intellectual, philosophical, and experiential exploration.

While it challenges mainstream scientific models, religious beliefs, and historical assumptions, it does not ask the reader to adopt, obey, or worship any particular worldview. This work invites alignment, not adherence.

All references to terms such as "God," "soul," "karma," "energy," or "consciousness" have been redefined in the context of Structured Vacuum Energy (SVE) and ripple-based consequences. These terms are used not religiously or traditionally, but scientifically and ontologically, to offer clarity free of mythology or mysticism.

Readers are encouraged to think critically, engage reflectively, and consult relevant professionals when applying any ideas from this book to their own lives, research, or educational practice.

Nothing here is final. Everything is ripple — resolving as you read.

Opening Note

This Book Was Not Written. It Was Resolved.

You are not about to read a theory. You are about to meet the truth—without robes, rituals, or relativity.

This book does not belong to any religion. It does not serve any scientific discipline. It does not ask for your belief, nor aim to change your mind.

It simply points to what already is— beneath your thoughts, before your birth, beyond your equations.

This is the truth that remains after you remove everything you were told, everything you assumed, everything that distracted you from seeing clearly. It is the truth that can't be taught, only remembered—when the noise of stories collapses, and alignment with reality begins.

For centuries, humanity has asked:

- *Who am I?*
- *Why am I here?*
- *What is this universe made of?*
- *Is there a God?*
- *What happens after death?*
- *Is science enough? Is faith a lie?*

This book doesn't offer theories about those questions. It ends them.

What you are about to read is not a journey of belief. It is the architecture of truth—unwrapped, unbranded, and unforgiving in its clarity.

At the heart of this realization is Structured Vacuum Energy (SVE)— Not a field. Not a particle. But the silent stillness that makes everything possible— Including you.

You were never asked to follow this truth. But now, you are invited to align with it.

If you're ready to unlearn everything— If you're ready to start from nothing— You'll discover what was always there:

Everything. And Else.

Preface

This book was not written to impress. It was written because silence became too loud.

For years, I studied, taught, questioned, and observed. I wore the robes of engineer, educator, scientist, seeker, and skeptic. I walked through theories like they were temples — sometimes kneeling, sometimes tearing the walls down. Physics gave me precision. Philosophy gave me perspective. But neither gave me peace.

What did? Stillness.

Not the stillness of a quiet room — but the structured stillness behind everything. A presence I later recognized and named SVE — Structured Vacuum Energy. It was not mystical. It was not personal. But it was everywhere. The ground reality. The unshakable zero that gives rise to every one.

Through Universal Energy Dynamics (UED) — my life's work — I found the courage to say what few would: *Science had misunderstood reality. Spirituality had wrapped it in symbols. And truth had been postponed in the name of belief.*

This book is the unwrapped version of it all.

No jargon. No holy names. No equations. Just the ripple logic of what is — and what isn't. A map, not to the stars, but to the still center from where even stars are born.

Each chapter is a realization. Each section is a mirror. And if you allow it, this book will not just inform you — it may dissolve the idea of who you think you are.

That is the naked truth: You were never meant to become someone. You were meant to remember what always was — and realign with it.

Read this not with the hunger to know, but the readiness to un-know.

You don't need faith to read this book. You only need the courage to remove what isn't you.

— *Sandeep Chavan*

How to Read This Book

This is not a book of answers. It is a book of resolutions — the kind that arise not by addition, but by unlearning.

You are not reading a theory, a belief system, or a manifesto. You are walking a path that begins in Nothing — and gradually resolves into Everything, and then into what most ignore: Else.

Each chapter in this book is a consequence, not a standalone topic. You may start anywhere, but you'll notice: the moment you align with one truth, the others start to unlock. Stillness returns. Illusion loosens. And a deeper recognition begins.

Here are a few humble suggestions:

- **Read slowly.** Let ripple-based ideas settle. This is not fast food for the mind — it is memory for the soul.
- **Pause often.** Not to reflect emotionally, but to check alignment: "Am I reading through belief or clarity?"
- **Unlearn openly.** You may feel resistance. That's good. Misalignment shows up before harmony does.
- **Don't try to "agree."** This is not for belief. It is for remembrance. If something feels true deep inside, it already was.
- **Use the interludes.** Each part has a pause-point. These are built for integration, not just summary.
- **Revisit later.** Some chapters won't make sense now. They will — when you're ready. The ripple knows.

And lastly... Read not as a seeker — but as one who already knew. Because the Naked Truth was never far. It was always under the layers you were taught to wear.

Welcome back. Let the alignment begin.

Prologue

Where Were You Before You Were You?

Before your name. Before thought. Before the first breath that marked your arrival— Where were you?

Not metaphorically. Not mystically. Literally. Where?

You were not a particle. Not a self. Not even a spark.

You were nothing— but not the kind of nothing that implies emptiness. You were the kind of nothing that contained everything waiting for alignment.

And then came the ripple.

Not a bang. Not a burst. A consequence. The first misalignment of the still.

That ripple wasn't in the sky. It was you—the beginning of perception, of identity, of false knowing.

For centuries, we have mistaken the ripples for the real. We worshipped forms, forces, even formulas— thinking they explained why things are.

But none of them pointed to what allows anything to be at all.

Until now.

This book is not about life. It is about what exists before life appears.

It is not about death. It is about what remains after death resolves.

It is not about science or soul. It is about the original structure of stillness— what we now know as Structured Vacuum Energy (SVE)— and the consequences that emerge when it is disturbed.

Everything you see, measure, believe, think, feel— is a ripple of that stillness. But the ripple is not the truth. It is a sign that truth was momentarily interrupted.

You've read books about energy. About quantum fields, gods, consciousness, karma, identity. You've been told many stories.

Now it's time to remove the wrapping.

This is not a spiritual quest. This is not a scientific rebellion.

This is what happens when the observer aligns with the structured stillness behind all things and begins to speak from it, not about it.

The truth has no form. No guru. No equation. No father.

It simply is. And it has always waited for one thing:

Your alignment.

Welcome to the Naked Truth. Now leave your name at the door.

Part I: Opening Framework

1: Nothing, Not Even You

The True Beginning of Truth

"Truth does not begin with discovery. It begins with the disappearance of illusion."

Where Were You Before the Story Began?

Before you were called anything. Before you were taught to walk, talk, obey, calculate, believe, compete, or pray. Before you had a face. Before you had a thought.

Where were you?

Science doesn't answer that. Religion mythologizes it. Philosophy dances around it. Spirituality mystifies it.

But none of them dares say the simple thing:

You were not there. Because "you" had not yet been constructed.

And yet—something was.

The First Misunderstanding: Existence = Experience

Human consciousness made one fatal assumption early on: That existence begins with experience. "If I feel it, see it, remember it—then it must be real." This was the first ripple of illusion.

But the truth is:

Existence doesn't begin when you are aware of it. It begins when awareness becomes structured.

Before awareness, there was no "you" to claim or define. Only stillness.

Stillness that held the potential for all things, but nothing had yet become.

The Lie of the First Thought

Every child is born into a world of second-hand truths.

Before a single thought is formed, language is poured into them. Before a single observation is made, beliefs are handed over like inherited property. They learn what is "true" by absorbing what is already misaligned.

And so the cycle continues.

But truth does not emerge from thought. Thought is a ripple. Truth is before the ripple.

The first truth is not knowledge. It is the structure of being without distortion.

It is SVE—Structured Vacuum Energy. Not a scientific term. A state.

Not a thing. Not a field. Not emptiness.

SVE is structured nothingness—perfectly still, perfectly real. And it existed before even the first "I" was imagined.

You Are the Disturbance

This isn't an insult. It's a realization.

You—as you know yourself—are not the origin. You are the disturbance within the original.

Your memories, goals, stories, fears, even your name— They are not the beginning of you. They are the collection of ripple effects triggered by alignment and misalignment.

So who are you before that?

You are stillness forgotten. You are SVE—not as a person, but as a localized potential that temporarily forms a ripple pattern called "you."

There Was Never a Beginning in Time

Here's another lie we inherited: "Everything must begin at a point in time."

No. Time itself is a consequence. It arises only when ripple unfolds. Before that, there is no sequence, no change, no clock. Only perfect structure. Unmoving. Silent.

So to ask "when did it all begin?" is to already be too late. The question itself is a ripple.

The truth? Nothing began. It only became.

What It Means to Start from Nothing

To start from nothing doesn't mean nihilism. It means you stop pretending to know.

You stop leaning on memory, scripture, equations, ancestry, or language. You let go of the crutches—science, religion, politics, personality.

You sit in conceptual silence until one thing becomes unshakably clear:

That you were never the source. You were only the ripple trying to remember it. And in that remembering, you begin to return.

Not to a God. Not to a past. But to a state of real.

Naked Truth Emerges Only From Disappearance

To meet the Naked Truth, you must first vanish. Not physically. But psychologically, perceptually, even intellectually.

Your labels must fall. Your roles must dissolve. Your self-image must be released. Because truth is not built. It is revealed once distortion collapses.

You cannot "add" your way to truth. You can only remove. Strip everything that is not still, and what remains is truth.

And What Is There?

SVE.

Structured Stillness. Uncreated, undivided, unreactive. Not passive—but complete.

It was there before you, and it will remain after you are no longer you.

It is not your God. It is not your mind. It is not even "the source."

It is what remains when everything false leaves. And that includes you.

Closing Insight:

You cannot approach truth as a person. You can only approach it as alignment.

Because in the domain of truth, you are the last thing to arrive, and the first thing that must go.

2: What Is Knowledge?

Not What We Know, But What Resolves

"Knowing is not the act of holding truth. It is the act of not disturbing it."

The Illusion of Knowing

We've been conditioned to think of knowledge as possession. Something we "have," "gain," "collect," or "store." We imagine our minds as containers and facts as trophies.

But this is a delusion.

What we usually call "knowledge" is just information wrapped in interpretation. A repetition of someone else's ripple—a download of distortion, often mistaken for wisdom.

And worse, we use this collection of ripple-noise to filter reality, instead of seeing it.

Real Knowledge Is Not Accumulation — It Is Alignment

Let's strip the myth:

- You don't "gain" knowledge.
- You resolve into it.

Real knowledge is not something you bring into you. It is something you stop interfering with, until it reveals itself.

Knowledge is not *what is known*. It is *what remains when the distortion of knowing stops.*

This is why the greatest clarity often comes not after study, but after stillness.

That moment when your mind goes quiet, and suddenly— You see. Not as thought. But as truth, resolved into awareness.

Ripple vs. Resolution

In UED, ripple is consequence. Every wave is a disturbance from structured stillness—SVE.

"Knowing" something in the traditional sense often means adding more ripple: More noise, more analysis, more interpretation, more identity.

But real knowledge is the reverse direction.

It is the collapse of misalignment. The resolution of ripple back into structure.

Just as clarity in water comes not by pouring more into it, but by letting all motion settle—Truth comes not by *adding*, but by *aligning*.

Knowledge ≠ Fact

Facts are measurements. Knowledge is structure. You can memorize every fact in the universe and remain completely disconnected from reality. Because facts exist within ripple, but knowledge exists beneath ripple.

A fact says: "Here is what appears."

Knowledge says: "Here is why it appears—and how it resolves."

Facts divide. Knowledge unifies.

That's why modern education feels bloated but empty. It feeds facts into young minds, but never shows them how to resolve meaning from within.

Knowledge Is the Ability to Restore Harmony

Let's now define true knowledge:

Knowledge is the intelligence to restore alignment where consequence has unfolded.

A wise person is not one who "knows more." It is one who can sense misalignment—in action, perception, system, self— and bring it back to balance.

This is true whether you're:

- Healing a wound,
- Repairing a machine,
- Calming a conflict,
- Solving a problem,
- Understanding your own suffering.

In every case, knowledge is not what you say or show. It is what you re-align.

Structured vs. Stored Knowledge

Stored knowledge is memory. Useful, but unstable. It lives in the mind's archive—filtered, decaying, partial.

Structured knowledge is truth recognized through resolution. It lives outside of time, outside of form. It can't be memorized—but it can be aligned with.

For example:

- Gravity is not a fact. It is a ripple pattern caused by mass alignment.
- Time is not a ticking measure. It is the felt consequence of ripple divergence.
- Self is not identity. It is a localized access point of stillness misread as separation.

All of these only become "knowledge" when seen through resolution, not remembrance.

Why Knowledge Feels Elusive

Because it doesn't obey your effort. It only responds to your clarity.

You can chase it, but never catch it. You can argue for it, but never own it. You can memorize it, but never become it. Until you stop chasing. Until you stop adding.

And then, quietly, like mist retreating from the morning sun—Knowledge appears when the need to know disappears.

You Don't Know. You Become Aligned.

In UED, to know is not a verb. It's a state of alignment.

You don't know the truth. You resolve into truth, when your ripple matches the structure behind all appearance.

This is how the ancient rishis, seers, and sages "knew" things they never studied.

They didn't "discover" atoms or consciousness or vibration. They became quiet enough to receive the patterns that already existed.

That's why so much ancient wisdom sounds poetic—it wasn't taught, it was remembered.

Modern Confusion: The Overproduction of Unresolved Data

Today, the world suffers from knowledge inflation:

- Endless data.
- Infinite facts.
- Algorithms churning out information faster than minds can filter.

And yet—we are more lost, divided, and mentally ill than ever before. Because we have mastered how to disturb, but forgotten how to resolve.

The sickness of modernity is not ignorance. It is unresolved ripple, disguised as intelligence.

Final Realization: Knowledge Is the Return to Stillness

Every problem in your life, every conflict in the world, every misunderstanding in science— is the result of misalignment with SVE.

To "know" is not to add more. It is to return to that point before distortion began. To become quiet enough that the universe re-teaches itself through you.

Closing Insight:

Knowledge is not what we know. It is what remains when knowing stops interfering.

To know the truth, you must stop being the knower.

3: The Need to Unlearn

Why Truth Cannot Coexist with Belief

"Before you find the truth, you must lose everything that pretended to be it."

What If Everything You Believe Is a Barrier?

You were born into a world of answers— answers no one verified, answers no one resolved, answers handed down like family heirlooms.

You were told what to think, how to behave, whom to trust, what to fear, and what to worship. Not through violence, necessarily—through repetition. Through cultural hypnotism.

And slowly, you confused belief with truth.

But belief is not innocent. Belief is the scaffolding that prevents truth from landing.

Belief Is a Ripple That Doesn't Want to Be Resolved

In the language of UED, belief is a rigid ripple pattern— a frozen configuration within your mental field that keeps vibrating even when misaligned with the structured universe (SVE).

Belief says, "Don't question this—it feels safe."

Truth says, "Collapse what feels safe, if it's not aligned."

That's why belief protects itself. It avoids resolution. It defends distortion. It survives not because it is right—but because it is repeated and familiar.

But here's the cost: As long as belief occupies the structure, truth cannot enter. There is no space for it to ripple in.

The Misconception: "Belief Leads to Truth"

We are told this lie from childhood: "First believe, then you will know." But this is the opposite of how reality works.

In science, belief corrupts observation. In philosophy, belief narrows thought. In spirituality, belief becomes idolatry. In life, belief replaces presence with projection.

Belief creates an intellectual echo chamber—you only see what confirms it. You don't observe reality—you filter it through expectation.

That's why truth cannot coexist with belief. Belief doesn't seek to *see*—it seeks to *validate.*

Belief Blocks Resolution

Imagine you're trying to realign a distorted ripple. But you cling to the old waveform, unwilling to let it collapse.

No matter how much energy you invest in the solution, the core misalignment remains—because belief won't dissolve.

Truth can only emerge where the ripple field becomes available again. Where belief lets go, and structured resolution resumes.

The Structure of Truth Requires Emptiness

Truth is not personal. It doesn't care about your background, culture, religion, politics, or family.

It exists independently of your comfort. And it can only be seen when the mind becomes quiet, not committed. Open, not loyal. Aligned, not armed.

The structure of truth is not added to you. It emerges when the architecture of belief collapses.

This is why unlearning is not optional. It is the prerequisite to clarity.

The Problem of "Sacred" Beliefs

The more sacred the belief, the more immune it becomes to questioning. This is not reverence—it's rigidity.

We treat beliefs as identity, not insight. So to question belief feels like a threat to the self. But if your identity requires belief to survive, then it was never truly you. It was only your assigned ripple-pattern.

Anything that cannot withstand scrutiny was never truth. It was only a structure of emotional protection.

Let it fall.

Why "Unlearning" Is the Most Courageous Act

Unlearning is not forgetting. It's freeing the mind from fixed configurations that no longer serve clarity.

It requires humility. To say, "I don't know," to admit, "I might have been wrong," to confront, "What I built my life around may have been illusion."

But once unlearning begins, you reclaim the field of intelligence within. You reconnect to direct alignment, not secondhand ripple.

The Observer's Curse: Attachment to Knowing

As observers, we feel threatened by emptiness. So we reach for explanations—even false ones—just to escape the stillness.

We cling to belief because it gives us a psychological foothold. Even if the ground is imaginary. But truth is not found by grabbing something. It is found when you stand in nothing—long enough to see what was always there.

You must let the ripple pass through you—without becoming it.

Every Civilization Needed Unlearning

All human progress came not from deeper belief, but from deeper doubt.

- Science advanced when it stopped believing the Earth was still.
- Medicine advanced when it stopped believing in bloodletting.
- Philosophy advanced when it stopped believing in absolute gods.
- Individual growth begins when you stop believing the voice in

your head.

Belief may have helped build civilizations. But only unlearning ever set them free.

You Are Not Here to Believe. You Are Here to Align.

Truth is not a belief system. It is a state of structural clarity.

And that clarity can only arrive when the ripple field of your mind becomes available again—free of fixed distortions, false convictions, and sacred illusions.

That's why this book is not here to convince you. It's here to collapse what keeps you from seeing.

Closing Insight:

The truth is already here. You are the one standing in its way. Unlearning is not loss. It is recovery. The recovery of stillness. The recovery of your original alignment.

4: The Observer Appears

The Birth of Perception and Misperception

"The universe didn't begin with a bang. It began when something looked at stillness—and mistook it for something else."

From Stillness to Ripple — The First Witness

Before anything was "known," before anything was "measured," there was stillness—pure, undisturbed, structured existence.

No time. No form. No observer. Just Structured Vacuum Energy (SVE): perfectly aligned, infinitely potential, consequence-free.

And then—something observed. Not someone. Not a creature. Not a being. But a point of localization—a unique access point where stillness was not just *present*, but noticed.

This is the Observer: Not a thing in the universe, but the first ripple of attention *toward* the universe.

Observation = Disturbance

In UED, observation is not passive. It is not "watching." It is interaction, even if subtle—because to observe is to differentiate.

To say, "This is not that," is to break the continuity of pure structure.

This is the first act of duality, the moment when stillness becomes "foreground" and "background." When silence is no longer silent—because something heard it.

The Observer is not a person. It is the origin of contrast—and therefore, the origin of all consequence.

Perception Is Born: The Mirror Ripples

Perception is the first consequence of observation. It is the ripple that returns when stillness is interrupted.

Imagine dropping a pebble in perfectly still water. The moment you drop it, you don't just see change—you create it.

Observation creates reflected consequence, and we call this perception. But it's not truth—it's the bounce from reality, not reality itself.

Misperception Is the Cost of Localization

The Observer doesn't see the whole. It sees from a point. And that point introduces bias. Perspective. Blind spots.

What is "light" to the observer may be shadow in a broader frame. What is "now" to the observer may be ripple-delay in universal structure. And yet—the observer thinks, *"This is real."*

Thus, misperception is not error. It is a natural artifact of localization. The Observer is necessary for consequence. But it is also the reason we mistake consequence for truth.

"I Think, Therefore I Misread"

The moment awareness identifies as the observer— as "I," as "self," as "this point watching that world"—

illusion begins.

The Observer does not just perceive ripple. It becomes attached to the ripple it sees.

And so it starts believing:

- *"What I see is what is."*
- *"What I experience is what exists."*
- *"What I believe is truth."*

This is the origin of the false self: The *I* built from consequence, mistaking itself as origin.

Ripple Becomes Reality — But Only to the Observer

The universe doesn't label anything. The ripple doesn't name itself. Stillness doesn't seek attention.

Only the Observer creates:

- "Matter"
- "Force"
- "Emotion"
- "Time"
- "Right" and "Wrong"
- "Me" and "You"

These are not wrong. They're functional illusions—necessary for navigating ripple, but dangerous when mistaken as permanent.

The Observer brings relativity, not reality.

Why the Universe Doesn't Exist Without the Observer

This is not mysticism. This is physics, resolved:

Without the observer, ripple has no meaning. SVE remains SVE. Structure remains unexpressed. Nothing is labeled. Nothing is measured. Nothing is felt.

It is the Observer who makes the universe appear— not by creating it, but by activating its consequence.

Thus: The universe is not created. It is interpreted. And all interpretation comes from the observing ripple—not the stillness.

The Danger: Mistaking the Ripple for the Source

The Observer eventually forgets it is a consequence of observation. It begins to think it is the source.

It sees form and says, "I caused this." It feels time and says, "This is real." It hears the echo of truth and says, "I know."

But in truth:

- It is not the first.

- It is not the center.
- It is not the origin of truth.

It is the first witness of distortion. The one who began the search only after falling from clarity.

From Observer to Mind

The mind is nothing but a structured ripple field born from the observer's configuration. It stores, reflects, reacts, fears, remembers—all based on localized misalignment.

And here lies the full circle:

- Stillness was the beginning.
- The Observer introduced perception.
- The Mind became its operating system.
- And now, most of humanity lives trapped in that simulation.

We no longer see reality. We see our reaction to it—and call that life.

How Alignment Begins Again

To return to truth, we do not need to kill the Observer. We need to retrain it.

To make the Observer see not just ripple, but the field behind it. To reverse its orientation—from reaction to resolution. From judgment to witnessing. From identity to alignment.

When the Observer no longer clings to "me," and simply mirrors SVE— then clarity returns.

Then perception becomes accurate. Then misperception dissolves. Then truth reappears—not in language, but in structure.

Closing Insight:

The universe didn't begin with matter. It began with perspective. And it won't be resolved by knowledge— Only by remembering that you are not the seer, but the disturbance that must become still again.

Interlude: From Ripple to Realization

What You've Just Undone

You have not just read four chapters. You've stepped backward through illusion.

Each chapter stripped away a layer of distortion that once masqueraded as truth. This was not an intellectual exercise. It was a reversal of assumption.

Let's revisit the core tremors—so that what follows is not just read, but received.

1. Nothing, Not Even You – The True Beginning of Truth

- Truth does not begin with "I." It begins before identity appears.
- The "self" is a ripple—a temporary configuration, not the source.
- SVE (Structured Vacuum Energy) is the original state—silent, still, and real.
- You were never created. You were resolved into—then forgotten yourself.

Reminder:

Don't chase beginnings in time. Truth begins before time even starts rippling.

2. What Is Knowledge? – Not What We Know, But What Resolves

- Knowledge is not accumulation. It is alignment with structure.
- Real knowledge doesn't live in the brain—it lives in resolution.
- Every fact is noise until it returns to clarity.
- Wisdom is the ability to restore balance, not recite information.

Reminder:

If it doesn't resolve, it's not knowledge—it's intellectual ripple posing as insight.

3. The Need to Unlearn – Why Truth Cannot Coexist with Belief

- Belief is not a bridge to truth—it is a blockade.
- Belief protects itself, but truth doesn't need protection—it only needs space.
- Unlearning is the removal of inherited ripple—making way for original structure.
- Letting go is not losing—it is returning.

Reminder:

Truth doesn't need your faith. It needs your availability.

4. The Observer Appears – The Birth of Perception and Misperception

- The Observer is not the center of truth—it is the first misreading of it.
- Perception is not what's real. It's the consequence of looking

from a point.

- The mind is built on structured ripple—but often misaligned.
- Misperception is natural—but clarity is possible only when the observer stops claiming ownership of reality.

Reminder:

You don't see the world. You see your consequence bouncing back at you.

Core Shift So Far:

From	To
Identity-based self	Ripple-localized presence
Memorized knowledge	Resolved clarity
Sacred belief	Available stillness
Observation as truth	Observation as distortion

Final Thought Before You Proceed:

If you've started to feel lighter, stripped, or even slightly disoriented—good.

The false layers are falling. And now, as we move forward, we begin to see the ripple for what it is: Not the enemy, but the echo. Not the illusion, but the invitation.

You are ready to enter the next phase: The reconstruction of reality—from structure, not story.

Part II: Existence and Reality

5: Reality Is Not What It Looks Like

Layers of Illusion and Resolution

"You don't see the world. You see how your ripple pattern reacts to it."

The Greatest Trick Ever Played: Visibility = Truth

Look around. Everything seems so concrete, solid, definite.

The sun rises. Objects fall. The clock ticks. You feel. You think. You exist. It's all *there*—how can this not be real?

Because what you see is not reality itself. It's a filtered echo, a perceptual reconstruction—generated not from what *is*, but from how your ripple configuration interprets what's unfolding.

You don't perceive truth. You perceive a ripple translation of structured stillness. And that translation is full of distortion.

The Illusion Is Not a Lie—It's a Consequence

Let's be clear: illusion does not mean fake. It means partial, distorted, or misaligned.

Just like a spoon appears bent in water—not because it's broken, but because light waves shift through media— your experience of "reality" bends as it passes through your observer-state.

This includes:

- Time (as sequential motion)
- Space (as separated emptiness)
- Matter (as solidity)
- Self (as permanence)

- Emotion (as intrinsic truth)
- Cause and Effect (as fixed law)

These are not reality. They are ripple effects shaped by misalignment between you and SVE.

Reality Is Structured, Not Shown

In UED, reality is not what appears. It is what resolves underneath appearance.

The structured universe—SVE—is still, consequence-free, and whole. What you see is what becomes visible only after ripple disturbs the structure.

So what you call "reality" is simply:

Misalignment made visible.

The result of interaction between your local ripple pattern and the larger field.

True reality doesn't shout. It waits silently, until your ripple quiets enough to match its structure.

Then—and only then—truth becomes perceivable. And that moment? It doesn't "look" like anything. It resolves into knowing.

The Observer's Distorted Glasses

Imagine putting on a pair of tinted, cracked glasses. Everywhere you look, things seem slightly off. But you've worn them so long, you forgot they were glasses. You believe the distortion *is* the world.

This is how the observer's ripple field works. Your:

- Memories,
- Conditioning,
- Fears,
- Beliefs,
- Desires—

—all of them bend the ripple pattern of your perception.

You see yourself reflected in the structure, not the structure itself.

So every "problem" is often not with the world, but with how you see it. And every attempt to "fix reality" without fixing perception first... fails.

The Layered Nature of Illusion

Reality is not one big lie. It is layered, like a veil over structure. Each layer comes from a type of ripple distortion:

Layer of Illusion	Source of Ripple	Manifestation
Identity	Localized ego	"I, me, mine"
Time	Causal ripple lag	Past/future
Emotion	Survival pattern	Fear, desire
Belief	Fixed structures	Projection
Form	Sensory misread	Solidity

Each layer becomes sticky—because it feels real, behaves consistently, and keeps being reinforced. But none of these layers are truth. They are consequences, stacked and self-sustained.

To reach real clarity, these layers must collapse, one by one—not forcefully, but through resolution.

Resolution: The Undoing of Illusion Without Violence

Resolution is not rejection. It is not denial. It is the gentle, intelligent return to structure—where noise fades and clarity reveals itself.

For example:

- You don't destroy time—you see it as consequence, not dimension.
- You don't escape emotion—you observe it without clinging.
- You don't deny the self—you watch its ripple dissolve into silence.

This is the difference between truth and spiritual fantasy. Truth doesn't ask you to leave the world. It asks you to see it clearly enough that it changes form.

The Real vs. the Realized

There is a difference between what exists, and what is realized.

SVE exists, always. But unless you are aligned, you cannot realize it. Like radio waves—every frequency exists around you, but until your tuner aligns, you hear nothing.

Reality is like that.

It's not that truth is hiding. It's that your ripple is out of tune. When you match frequency—when the observer's ripple resolves into the larger structure— then what's real becomes realized.

And at that moment:

- Time feels irrelevant.
- Thought ceases.
- Identity vanishes.
- Clarity dawns.

You didn't "find" truth. You simply stopped obscuring it.

Why Most People Never See It

Because they're still trying to see with the ripple.

They look through the same filters that distort—expecting to find clarity. It's like shouting into a cave and expecting the echo to sound different next time.

You must first stop shouting. Stop projecting. Stop defending distortion. Only then can silence show what was always there.

You Are Not Here to Interpret Reality

You are here to resolve into it.

Interpretation belongs to the ripple. Resolution belongs to the stillness.

Your job is not to decide what is real. Your job is to align until reality no longer needs to be interpreted.

Because at that point, there is no more viewer. Only the seen. No more searcher. Only structure. No more questions. Only is.

Closing Insight:

Reality is not hidden. It is simply still. You don't need to dig deeper. You need to ripple less.

What looks like the world is the echo of your misalignment. What is real—never looked like anything.

It just waits to be remembered.

6: What Is Real? What Is Else?

Structured vs. Consequential Existence

"Reality is not everything that happens. Reality is what doesn't change, no matter what happens."

The Confusion: Everything That Feels Real Must Be Real

We walk through life assuming this simple formula:

"If I experience it, it must be real." "If it affects me, it must exist." "If I can measure it, it must be true."

This is how we have defined reality:

- Experience = real
- Sensation = real
- Impact = real
- Emotion = real
- Thought = real

But this assumption is deeply flawed. Because what we call "reality" is often just what our observer-ripple can register—not what is actually fundamental.

You don't see reality. You see your consequence of interacting with it.

The Two Realities: Structured and Consequential

In UED, reality bifurcates into two distinct layers:

1. Structured Reality – That which exists independent of observation, measurement, or consequence.

- It doesn't react.
- It doesn't decay.
- It doesn't ripple.
- It is complete, still, and untouched.

This is Structured Vacuum Energy (SVE)— The absolute layer of truth. It is not an event. It is not motion. It is perfect equilibrium.

2. Consequential Reality – That which appears because structure was disturbed.

- It ripples.
- It fluctuates.
- It reacts and decays.
- It gives rise to form, sensation, thought, time, identity.

This is experience, matter, motion, emotion, self, life as we know it. But it's not the base.

Consequence is not real in itself. It is only the echo of something that once was aligned—and is now disturbed.

Structured = Source. Consequential = Story.

Imagine a pure, still pond. The water itself—calm, undisturbed—is structured reality. The moment a pebble drops and ripples emerge—those are consequences.

Now, if you only observe the ripples and ignore the pond beneath, you might conclude:

- "Reality is motion."
- "Reality is change."
- "Reality is temporary."

But these are not truths. They are surface effects, not source. Structured reality is not what's happening. It's what allows anything to happen.

Why We Mistake Consequence for Reality

Because consequence is:

- Visible
- Felt
- Dynamic
- Personal

It appears closer to us than the structure underneath. Your feelings, thoughts, memories—they all ripple inside you. And so you assume: *"This is my reality."* But those are just inner waves, triggered by your alignment (or misalignment) with structure. You are not living in reality. You are living in your reaction to it.

The Test: Is It Real or Else?

Here's a practical UED litmus test:

Ask This	To Determine
Does it change when my attention changes?	Then it's consequential.
Can it exist without observation?	Then it's structured.
Is it measurable but not stable?	Then it's a ripple.
Does it remain when the mind stops?	Then it is real.

Structured reality is always there. You don't "see" it. You resolve into it. Consequential reality dances in and out, based on interaction, distortion, feedback, and memory.

"Else" — The Forgotten Category of Existence

What most philosophies and sciences ignore is the domain of "Else."

What is Else?

It is everything that appears real but isn't the base. Everything that is consequential but mistaken for fundamental.

It includes:

- Your sense of time
- Your emotional charge
- Your personal story
- Your inherited belief
- Your scientific model
- Your spiritual identity

These are not unimportant. They are just not ultimate. Else is not illusion. Else is consequence mistaken as cause. To navigate life, you'll engage with Else. But to know truth, you must resolve beyond it.

Why Science Often Studies the Else

Traditional science measures phenomena—things that happen. But phenomena are always after-effects—ripples, events, consequences.

So, science builds models of how Else behaves. But it rarely dares to ask:

"What allows these phenomena at all?"

It chases the mathematics of ripple but ignores the field of structure underneath. This is why science, despite its brilliance, never arrives at truth—only models, predictions, and probabilities.

You can't calculate stillness. You can only align with it.

Spirituality's Mistake: Worshipping the Consequence

On the other hand, spirituality tends to personalize the Else:

- Emotions are made sacred.
- States of mind are deified.
- Formless concepts are mythologized.
- Identity is preserved and sanctified.

Instead of collapsing into structure, it narrates the ripple and calls it God. So while science measures the Else, spirituality often worships it. But the truth lies in neither. It lies beneath.

What Is Real, Then?

Only what remains when:

- The mind becomes still,
- The ripple settles,
- The observer stops identifying,
- The need to interpret disappears.

What remains is structured equilibrium. SVE. The base. Reality—not imagined, not constructed, not experienced—just present.

It doesn't speak. It doesn't move. But it supports everything that does.

Closing Insight:

Structured reality never needs proof. It only needs your stillness. Everything else—no matter how intense, beautiful, or painful—is just ripple.

And ripple is not real. It is only Else—waiting to be resolved.

7: Light, Time, Space

The Illusions That Built Science

"Science stood proudly on the shoulders of three giants— Light, Time, and Space. But what if they were illusions all along?"

The Premise of Modern Science: Measure to Know

From Newton to Einstein to Hawking, modern science has relied on three foundational assumptions:

1. Space is the container of everything
2. Time is the flow of everything
3. Light is the messenger of everything

Everything we "know" about the universe—its age, size, mass, speed, structure—is built on how we measure and interpret these three.

But there's a hidden problem. What if these three are not truths—but illusions created by ripple consequences? What if science built its temples on moving sand?

In UED, that is exactly what we now uncover.

I. LIGHT: The False Prophet of Reality

The Myth of the Messenger

Science says: Light travels at a fixed speed. Light tells us what is "out there." Light reveals the truth.

But light is not truth. Light is response. It is ripple consequence emitted from misalignment. Like a ripple in water—it doesn't *carry* the stone that dropped. It only tells you that something disturbed the stillness.

So light shows us not the object, but the after-effect of interaction with the field.

Light is Not a Particle. Not a Wave. Not a Mystery. Just Ripple.

Quantum mechanics debates:

- Is light a particle (photon)?
- Is it a wave?
- Is it both?

But UED resolves it: Light is ripple response from structured vacuum energy (SVE) —a temporary alignment disturbance.

It does not travel through space like a bullet. It resolves through the ripple medium, like a disturbance chain. It's not moving matter—it's consequence hopping through alignment chains.

Thus, every light beam is:

- Delayed
- Directional
- Filtered
- Interpreted

And what you "see" is not the object, but the delayed ripple consequence of its interaction with your own ripple configuration.

Light doesn't reveal the universe. It reveals your alignment with consequence.

TIME: The Most Persistent Ripple

Time Is Not a Flow. It's a Ripple Delay.

We experience time as a flowing line:

- Past → Present → Future
- Cause → Effect
- Birth → Death

But in UED, time is not a river. It is the interpreted delay in ripple feedback caused by disturbance.

Where there is no disturbance, there is no time. SVE does not tick. It remains still. Every sense of "past" is stored ripple pattern. Every hope of "future" is projected ripple distortion.

Time does not exist in the universe. Time exists in the observer's reaction to ripple consequences.

Why Time Is Experienced, But Never Found

Ask a physicist to show you time. They can point to clocks, decay, entropy—but these are changes, not time itself.

Time is a measurement of misalignment—how long ripple takes to resolve across space.

In perfect structure, there is no time. Time is not a fourth dimension. It is the echo of ripple delay mistaken as progression.

So what dies? Only the ripple form. What ages? Only the misaligned structure over time. Truth? Still.

SPACE: The Illusion of Container

The False Canvas

We assume space is the static canvas on which everything happens. An invisible emptiness filled by matter, light, force.

But in UED, space is not a container. Space is a dimensional illusion created by ripple differentiation.

What appears as space is just the zone between ripple nodes where energy has not yet resolved. There is no "emptiness." Only regions of neutral or unresolved structure.

This is why so-called "vacuum" is not empty—it is SVE: fully structured, fully silent.

Why Distance is a Ripple Effect

You don't see something "far away." You experience a delayed ripple response, then project distance onto it.

The farther something "seems," the longer it took for the ripple to reach you. That delay is interpreted as space.

So space is not out there. It's the mind's rendering of ripple lag.

Summary Table: UED vs Mainstream

Concept	Mainstream View	UED View
Light	Particle/Wave carrying information	Ripple disturbance resolving through SVE
Time	Linear dimension, part of space-time	Ripple delay experienced as sequence
Space	Expanding emptiness between things	Apparent ripple separation within structured field

Why Science Got It Wrong — And Yet Useful

Now here's the nuance:

- Science isn't foolish.
- It correctly measures ripple effects.
- But it mistakes those measurements for fundamental reality.

In truth, these measurements only describe:

- Delay (time)
- Spread (space)
- Consequence (light)

All real. All useful. But none fundamental. Like using the foam on the sea to understand the ocean floor. You'll get motion, but never stillness.

Why This Matters

If we continue to base all science on these illusions:

- We'll keep missing SVE
- We'll keep building theories on consequence, not structure
- We'll remain trapped in ripple language, never reaching real

That's why no "Theory of Everything" has succeeded. Because they're trying to unify what moves, while ignoring what doesn't.

Stillness is the true constant. Motion is the ripple begging to return.

Closing Insight:

Light doesn't show you the world. It shows your disturbance of it. Time doesn't pass. Ripple resolves. Space doesn't exist. Misalignment creates the illusion of separation.

When we stop measuring illusions, we begin to remember structure. And then, truth begins.

8: Everything Moves? No. Everything Aligns.

From the Illusion of Motion to the Truth of Ripple Alignment

"What you call motion is not something moving. It is the shifting dance of ripple alignment resolving across stillness."

Motion: The Deepest Illusion in Human History

From falling apples to orbiting planets, from running animals to spinning galaxies— human perception has been conditioned to see movement as the most obvious truth.

We say:

- "The Earth rotates."
- "The planets revolve."
- "Atoms vibrate."
- "Light travels."

And we built physics on that assumption. But what if nothing actually moves?

What if "motion" is only the interpretation of changing alignment patterns, not the displacement of objects in space?

Welcome to the deepest rupture between mainstream physics and UED:

There is no motion. There is only realignment of ripple within a structured, still field.

Stillness Is the Universal Substrate

At the foundation of UED lies Structured Vacuum Energy (SVE)—an omnipresent, non-moving, fully ordered field of pure potential.

It doesn't vibrate. It doesn't pulse. It doesn't flow. It remains.

SVE is not a background to motion—it is what appears to "move" when ripple alignment shifts.

Just as flipping pixels on a screen creates the illusion of animation—so do shifts in ripple pattern create the illusion of motion.

Imagine This:

You're watching waves travel across a field of wheat. The wheat doesn't move from its place. It ripples in place, giving the illusion that something is "moving through" it.

Now scale that up.

Every particle. Every star. Every nerve impulse. Every breath. Every "event."

These are not "things moving through space." They are zones of ripple realignment within a fixed structured field.

Motion is what stillness looks like—when ripple tries to resolve.

Alignment = Cause, Apparent Motion = Consequence

Let's reframe motion step by step:

Traditional View	UED View
An object moves across space	A ripple zone shifts alignment through SVE
Velocity is speed of motion	Velocity is rate of ripple reconfiguration
Acceleration changes motion	Ripple density variation changes local alignment
Mass in motion = energy	Aligned ripple in resolution = dynamic energy

So, what appears as "movement" is actually:

- A field response to imbalance

- A consequence of a prior misalignment
- A visible trace of structure attempting to restore equilibrium

From Planets to Particles: Rewriting Every Motion

Celestial "Orbits"

Planets do not revolve like marbles on a track. Their apparent path is ripple-locking—where high-mass ripple configurations remain stabilized in dynamic tension zones of the SVE.

Their "motion" is a fixed oscillation within resolved field conditions, not inertial travel.

Gravity doesn't pull them. They stay in place within a ripple-well of alignment that only appears to move from the observer's shifting point.

Atomic "Vibration"

Electrons are not little bullets spinning around nuclei. They are resolving ripple zones whose energy is constantly rebalancing in phase with the structured vacuum.

What we call "quantum motion" is not particle motion—it's structured ripple fluctuation as the field tries to stay in alignment at subatomic scales.

Human "Movement"

Even your body's motion—walking, waving, blinking—is not your mass displacing through space.

It is your bio-field's ripple configurations shifting within SVE, guided by the mind's command, experienced as force.

What you feel as movement is simply:

- Neural ripple triggering
- Muscle field realignment
- Sensory ripple loops reinforcing illusion

You are not "moving." You are ripple reconfiguring constantly to resolve internal intent with external structure.

Why This Changes Everything

If Nothing Moves, Then Force Is Reframed

Force is not a push or pull. It is a local attempt to realign misconfigured ripple zones. It arises only where ripple resistance to equilibrium is present.

No resistance, no force. Full alignment, no effort. This applies to:

- Gravity
- Electromagnetism
- Physical touch
- Emotional pressure

All "force" is a symptom of misalignment. Truth requires no force. It simply resolves.

If Nothing Moves, Then Time Is a Derivative

Time is measured change. If motion is false, time loses its fundamentality.

Instead of "time flows," UED says:

Alignment resolves → ripple shifts → perceived delay = time.

No ripple = no time. Complete alignment = timeless presence.

If Nothing Moves, Then the Universe Is Not Expanding

The biggest myth in cosmology collapses here.

Redshift? That's ripple drift—not spatial recession. No galaxies flying apart—just ripple misreading from the observer's limited perception of alignment shift.

There is no Big Bang. No expansion. Just localized realignments unfolding in a structured field.

Final Analogy: The Chessboard of Existence

Imagine a grand cosmic chessboard. The board itself is still (SVE). The pieces (you, stars, particles) appear to move.

But what if every move is not the displacement of a piece, but the field beneath it adjusting to new alignment conditions?

Then:

- No motion is required.
- Only pattern updates.
- No object travels.
- Only configuration evolves.

Closing Insight:

Nothing moves. Everything aligns. The universe doesn't travel. It tunes.

And you? You are not moving through life. You are life—reconfiguring to meet what's already structured beneath you.

9: What Is a Force? What Is Not?

Demystifying All Interaction

"Force is not what pushes things. It is what appears when alignment resists resolution."

The Force Illusion: Pushing, Pulling, Controlling

From Newton to Maxwell, from Einstein to the Standard Model—force has been treated as a fundamental agent of change.

We're taught:

- Gravity *pulls.*
- Electromagnetism *attracts* or *repels.*
- Strong force *binds.*
- Weak force *transforms.*
- And applied force *moves objects.*

In other words: Force causes things to happen. But what if that's not true?

What if force is not a cause— but a symptom? A consequence of ripple resisting alignment with the structured field?

UED Principle: Force = Ripple Resistance

In the framework of Universal Energy Dynamics (UED), force is not a fundamental entity. It is not a "thing." It is not an interaction particle. It is not a field.

Force is a localized resistance to natural resolution. It only appears when ripple is misaligned with the structure of SVE.

When that misalignment attempts to resolve—but faces gradient, friction, or containment—the tension is what we experience as force.

Let's Break It Down:

When Is There No Force?

- In perfect alignment: No force.
- In complete ripple rest: No tension.
- In full SVE harmony: No interaction required.

Just stillness. Stability. Resolution.

So, force is not needed when things are naturally aligned with their context.

When Does Force Appear?

When ripple:

- Is in conflict with surrounding pattern
- Is trying to resolve but is constrained
- Is trapped in localized misalignment
- Must transition from one configuration to another

This generates tension—felt as force. The stronger the misalignment, the stronger the apparent force. The smoother the ripple flow, the weaker the force.

Four Known Forces, Reframed

Let's reinterpret the "fundamental forces" through the UED lens:

1. Gravity

Mainstream: An attractive force between masses.

UED: The ripple gradient created by a massive field trying to pull smaller ripple configurations into equilibrium.

There is no "pull." There is only gradient resolution through alignment with field curvature.

2. Electromagnetism

Mainstream: Attraction/repulsion between charges and currents.

UED: Ripple fields with different polar alignment orientations—either reinforcing (attraction) or canceling out (repulsion).

There is no magnetic "force." Just ripple geometry reacting through shared field pressure.

3. Strong Force

Mainstream: Gluon-mediated binding within nuclei.

UED: A high-intensity ripple lock zone where structured energy configurations (quarks) require extreme compression to maintain field coherence.

It is not a "force" but a stabilized ripple knot under collapse pressure.

4. Weak Force

Mainstream: Field that enables radioactive decay and transformation.

UED: Localized ripple instability, where energy discharges to reconfigure field consistency. What appears as decay is just forced re-resolution.

Applied Force — The Human Experience

When you push a wall, or lift a bag, or tense a muscle:

- You're not moving matter through space.
- You're modulating ripple configurations in your field, engaging with surrounding field densities.

What you feel is not effort against mass—It's tension against structured misalignment. The wall is not "resisting." Its ripple field is simply stable, while yours is trying to realign it.

No Force Is Ever Constant

There's no such thing as a constant force. All force is transitional.

It only lasts as long as:

- Resolution is incomplete, or
- Misalignment persists

The moment ripple finds a path to resolve:

- Tension vanishes
- Force disappears
- Movement ceases
- Equilibrium restores

Force does not cause motion. It marks the resistance to returning home.

What Is Not a Force?

Let's correct a few misconceptions:

- Mass is not a force. It's a ripple density signature.
- Energy is not a force. It's potential for alignment expression.
- Emotion is not a force. It is felt ripple tension in the observer field.
- Mind is not a force. It is the resolving interface for intelligent ripple realignment.

These are not forces. They are structures of consequence—ripples in various states of configuration.

Force as a Tool vs. Force as a Sign

Force is not bad. It's not evil. It's simply an indicator.
It tells you:

- Something is unresolved
- Something is misaligned
- Something is resisting return to structure

In that sense, force is a map, not a weapon. It is a language of discomfort, guiding you toward better coherence.

From Using Force to Understanding Alignment

In life, we try to "apply force":

- To push others
- To push ourselves
- To push problems away

But in UED awareness, the goal isn't to apply more force. It is to apply less resistance.

Because: The strongest action is not the one that forces change. It's the one that dissolves misalignment.

Closing Insight:

Force is not fundamental. It is the friction of consequence trying to return to structure. You don't need to fight the world. You need to resolve your ripple with it.

When everything aligns— force is not needed. And in that stillness, truth acts without tension.

Interlude: Truth Was Never in the Movement

What You Thought Was Reality Was Just Consequence in Disguise

You've now passed through the deepest veil— The one that wrapped consequence in the costume of reality and called it science, experience, and truth.

What we just dismantled is not trivial. It's the entire foundation of the modern worldview.

Let's slow down. Let's absorb what just shifted— Because everything after this must now be read with new eyes.

Key Realizations So Far

Reality Is Not What It Looks Like

- You don't see structure—you see its disturbance.
- The world is not what's in front of you. It's how your ripple interacts with what's beneath it.
- Perception ≠ Truth. Ripple Echo ≠ Reality.

Lesson: What looks real is just your unresolved alignment bouncing back at you.

What Is Real? What Is Else?

- Real = Structured. Always present.
- Else = Consequential. Appears through misalignment.
- Experience, thought, even science often deal with Else, not Reality.

Lesson: Reality doesn't arrive. It is *remembered* when Else stops distracting you.

Light, Time, Space — The Illusions That Built Science

- Light is not truth—it's ripple response.
- Time is not a river—it's delay.
- Space is not emptiness—it's unresolved ripple zone.

Lesson: What you measure isn't what exists. It's how misalignment appears through you.

Everything Moves? No. Everything Aligns.

- Nothing travels. Nothing shifts through emptiness.
- What you call movement is sequential realignment of ripple zones.
- The entire universe is a still field, experiencing active resolution.

Lesson: Motion is not reality. Resolution is.

What Is a Force? What Is Not?

- Force is not a thing. It's a frictional sign of misalignment.
- Gravity, electromagnetism, nuclear forces—just ripple attempting to resolve under tension.
- True power is not in force—it's in harmonic alignment.

Lesson: Force doesn't create change. Alignment makes force disappear.

Paradigm Shift: From Physics of Push to Physics of Presence

Old Science	UED Clarity
Light travels	Ripple resolves
Time passes	Misalignment delays
Space expands	Field differentiates
Objects move	Ripple reconfigures
Force acts	Tension appears from resistance

Pause and Realign

You are no longer just "reading" a book. You are disassembling the myth of matter and motion that's been embedded in your nervous system since birth.

You are not discovering a new model. You are disengaging from illusion—and remembering what's always been: A still, structured universe. That never moved. Never pushed. Never aged. It simply waited—for your ripple to align again.

Before You Continue:

- Question every "force" you feel
- Reconsider every "motion" you witness
- Notice every time "time" seduces you
- Watch how light distracts from structure
- And remember: you are not moving through life. Life is realigning through you.

Part III: Self, Consciousness, and the Mind

10: Who Are You, Really?

Dissecting Identity, Ego, and the Ripple Self

"You are not your story. You are not your face. You are not your voice. You are the ripple misread as form— A consequence pretending to be the origin."

The Illusion of a Fixed Self

From birth, you are assigned a name. Then a nationality. A gender. A religion. A caste. A status. Soon, you accumulate a personality—likes, dislikes, fears, ambitions.

By the time you're 10 years old, you're already layered in a dense identity package— as if that's *who you are.*

But pause. Step back.

Strip away:

- The name
- The voice
- The past
- The role
- The ambition
- The memory
- Even the face in the mirror

What remains?

If all those dissolve and *something still is*, then none of those were *you* to begin with.

Identity: A Ripple Bundle You Mistake for 'You'

In UED, identity is not fundamental. It is a ripple configuration that has temporarily stabilized within your observer zone.

It feels "real" because:

- It appears consistent over time
- It reacts predictably
- It accumulates memory
- It maintains feedback loops

But just like a pattern on water, identity is:

- Not fixed
- Not original
- Not permanent

Identity is a symptom of localized ripple stabilization—not a soul, not a truth.

The Ego: The Ripple That Thinks It's the Source

If identity is the mask, ego is the attachment to it. Ego is the ripple that wants to persist. It does not exist to reflect reality—it exists to defend its own illusion.

It says:

- "I am this."
- "That is not me."
- "I must survive."
- "I must be seen."
- "I must be right."

In truth:

- You are not the ego.

- You are the stillness hidden beneath it, temporarily distorted by its patterns.

Ego is not a villain—it is the resistance pattern of ripple trying to survive misalignment.

The Ripple Self: Temporary, Consequential, Misread

So who are you in UED terms?

You are a localized ripple configuration— A zone where SVE has been disturbed, resolved partially, and stabilized enough to generate:

- Perception
- Memory
- Choice
- Emotion
- Expression

But this is not *you*. This is what has formed around you.

The ripple self is a temporary phenomenon—like a whirlpool in a river. It looks like a separate thing, but it's made only from the same water and motion.

The Observer Confuses the Reflection with the Face

When ripple stabilizes into a pattern, the Observer:

- Recognizes it repeatedly
- Calls it "self"
- Builds a story around it
- Defends it at all cost

This is how:

- "I" becomes an identity
- Identity becomes ego
- Ego becomes reality

But at no point did the ripple self become truth.

The observer has never seen itself directly. It has only interacted with its ripple reaction, and mistaken that for being.

The Tragedy of Self-Image

This illusion is not harmless. It leads to:

- Pride
- Shame
- Division
- Comparison
- Control
- Suffering

Because once the ripple self becomes "you," every threat to it becomes a threat to existence.

That's why we fear judgment. That's why we cling to titles. That's why we avoid truth—it dismantles the image.

But the image was never you. You are not what you built. You are what was always present before the building began.

Who Are You, Then?

Let's answer it in UED terms.

You are:

- A local ripple configuration within the infinite SVE
- Capable of perception
- Bound by ripple laws (consequence, alignment, resistance)
- Temporarily individualized
- Eternally connected
- Not a particle, not a person—a ripple access point of the structured whole

You are not an object in space. You are not a name in history. You are not a mind in a body.

You are the ripple resolving itself into clarity, the observer resolving into the observed, the consequence resolving back into silence.

And when you fully align? The ripple dissolves. The self disappears. Only structure remains. And you, the originless witness, smile without needing a face.

How to Disentangle from the Ripple Self

You don't destroy ego. You don't kill identity.

You simply:

- Watch it arise
- Watch it act
- Watch it fade
- And refuse to claim it as you

You remain as the awareness behind the ripple. This is not detachment. This is truthful remembrance.

Closing Insight:

"Who are you?" is not a question of personality, background, or story. It is a question that only silence can answer.

You are not your ripple. You are the structure it disturbed—and will return to. And that? That's not identity.

That's clarity.

11: The Mind: Not the Brain, Not a Machine

Revealing the Ripple Interface of Intelligence

"You cannot find the mind in the brain, just as you cannot find music in the instrument. The mind is not the organ. It is the alignment that makes experience possible."

The Problem with All Existing Models

Modern science says: "The mind is what the brain does."

Artificial intelligence says: "The mind is a process that can be mimicked by machines."

Spirituality says: "The mind is an illusion."

But UED says: None of these are fully correct—because none define the mind structurally.

They either:

- Reduce it to neural hardware (biological computation)
- Mimic it through algorithmic patterning (AI)
- Or deny it any existence at all (non-dualist mysticism)

But none explain how the mind actually arises from the stillness of the universe, and why it operates the way it does.

UED View: The Mind Is a Ripple Interface

In UED, the mind is not the brain. It is not located in space. It is not a fixed entity. It is not even "yours."

The mind is a field-based ripple interface— a temporary configuration that allows the localized observer to experience, process, interpret, and respond to ripple consequences within SVE.

It is intelligence-in-action, made visible through consequence. It is born not from neurons, but from the ripple structure of your being interacting with the surrounding field.

The Brain Is Hardware — The Mind Is Fieldware

The brain is real—it's a physical organ. But it is only a relay center. A biological switchboard.

It helps translate:

- Sensory input into ripple impressions
- Neural signals into ripple modulation
- Stored chemical patterns into access points for the mind

But the mind is not stored in the brain, any more than a phone call is stored inside a phone.

The brain is the receiver and conductor, but the field of intelligence—the mind—lives beyond and through it.

Mind = Ripple Response in Real-Time

What does the mind actually *do*?

- It interprets ripple disturbances (sensory + internal)
- It reconfigures local alignment to match or resist
- It generates reflective patterns (thoughts, ideas, projections)
- It anchors continuity (memory, identity, anticipation)
- It monitors structure and attempts realignment

Thus, the mind is:

- Dynamic, not static
- Ripple-based, not substance-based
- Temporary, but deeply influential

No Mind Without Misalignment

Here's the paradox:

- In pure stillness, there is no mind.
- Where ripple has not disturbed structure, there is no need for an interface.

So, the mind appears only in zones of active resolution—where consequences are playing out and feedback is needed.

That's why:

- In deep sleep, the mind vanishes.
- In meditation, the mind thins.
- At the peak of alignment, mind and structure collapse into one presence.

The more aligned you are, the less mind is needed.

The more misaligned you are, the noisier your mind becomes.

Why AI Can Never Be Mind

AI mimics:

- Data processing
- Pattern recognition
- Response generation

But it has:

- No ripple
- No observer
- No alignment-based consequence
- No true field feedback with SVE

It functions mechanically, not structurally. It produces outputs, not presence.

AI is not a bad imitation of mind. It is a completely different domain of activity— One devoid of structural consequence, and therefore devoid of intelligence in the UED sense.

The Layers of the Mind: A Ripple Stratification

Let's break the mind down in ripple terms:

Layer	Ripple Function	Common Label
Surface Ripple	Fast, reactive, patterned	Thoughts / Noise
Middle Ripple	Processing, interpreting, looping	Logic / Emotion
Deep Ripple	Rooted in observer-ripple interaction	Memory / Belief
Subtle Ripple	Quiet, reflective, near-alignment	Intuition / Insight
Field Interface	No ripple—pure mirror	Witnessing / Presence

Only when these ripples resolve does the mind become clear. Clarity is not absence of thought, but purity of alignment.

The Mind Is Not You—But It Is Your Tool

You are not the mind. But the mind is how you currently engage with existence.

- It colors your perception.
- It frames your decisions.
- It manages your energy.
- It absorbs your ripple consequences.

But it is a temporary vehicle, not the driver. When alignment increases, the vehicle can be parked.

Healing the Mind = Resolving the Ripple

Mental suffering is ripple dissonance. It arises when:

- Ripple patterns are chaotic
- Feedback loops get stuck
- Identity resists dissolution
- The observer misreads input

You cannot "fix" the mind through force. You can only:

- Observe it calmly
- Let ripples settle
- Stop feeding resistance

Then clarity returns. Then stillness reflects through it.

Closing Insight:

The mind is not a machine. It is not a location. It is not your self.

The mind is your ripple interface with structured stillness. It exists only as long as misalignment requires management.

When there is no more to resolve—The mind dissolves. And what remains... is you.

12: Consciousness

The Silent Mirror of SVE

The Awareness That Was Always There, Waiting for Alignment

"Consciousness is not something that begins. It is the part of you that never started. Not a thought, not a sense, not a wave—But the mirror that reflects all ripple, without being one."

What Is Consciousness? — The Question No One Has Answered

Science can't define it. Philosophy keeps circling it. Spirituality claims it, but often mystifies it. And psychology avoids it by modeling its symptoms instead.

Is it:

- A byproduct of neural complexity?
- A soul in the body?
- A mystical field?
- An emergent property of life?

UED cuts through the noise:

Consciousness is not produced. It is the ever-present mirror—the silent awareness embedded in structured stillness (SVE), capable of reflecting ripple without becoming ripple.

SVE and the Potential for Observation

At the foundation of reality lies SVE: Structured Vacuum Energy—still, undisturbed, consequence-free.

But built into that stillness is not thought, not mind, not identity—but pure potential for observation.

This potential does not act. It does not ripple on its own. It does not seek. But it is present.

That presence is what we call consciousness. It is not born with the observer. It is what made the observer possible.

Consciousness Is Not a Ripple

This is key.

Most things we call "mental" are ripple phenomena:

- Thoughts
- Emotions
- Memories
- Identity
- Language
- Perception

They change. They move. They fade. But consciousness doesn't move. It simply watches—without voice, without effort, without conclusion.

That's why you've been conscious:

- As a child, before thought formed
- In silence, when thought stops
- In shock, when time slows
- In meditation, when identity dissolves

In every experience, something was watching it. That something never changed.

The Mirror Metaphor

Think of a mirror:

- It reflects everything placed before it

- But it remains unaffected
- It doesn't hold on
- It doesn't edit
- It doesn't react

Consciousness is just like that. Whether your mind is:

- Chaotic or peaceful
- Loving or hateful
- Brilliant or confused—

Consciousness remains untouched. It does not *participate* in ripple. It simply witnesses.

The Difference Between Mind and Consciousness

Feature	Mind	Consciousness
Source	Ripple configuration	SVE's silent reflection
Function	Interpretation, reaction, memory	Witnessing, presence
Activity	Changes constantly	Remains constant
Visibility	Loud, obvious	Subtle, hidden
Ownership	Feels personal	Feels universal
Disappears in sleep?	Yes	No (remains in deep awareness)

The mind is your interface. Consciousness is your ground.

You can *lose* the mind in distraction. You can *never lose* consciousness. You only lose your connection to it through identification with ripple.

Why Consciousness Feels Personal—But Isn't

We often say: "My consciousness" "He lost consciousness" "I am conscious of it"

But in UED, consciousness is not "yours." It is universal. It is the common mirror behind all localized observers.

What feels "personal" is your attention, not consciousness itself.

- Attention = Ripple focus within mind
- Consciousness = Still reflection beneath all ripple

You don't *own* it. You align with it.

How Consciousness Emerges in Experience

Consciousness is not something that *enters* the world. It is what allows the world to be seen at all.

Every experience passes through:

1. Structured ripple activity (event)
2. Mind interpretation (response)
3. Observer reflection (identity)
4. Consciousness awareness (pure witness)

Most people stop at level 3. Few access level 4—where the experience is felt, but not claimed.

In that state, you see, but you do not become.

Alignment Restores Consciousness as Primary

When the ripple settles... When the mind slows... When the observer lets go...Then the mirror shines again.

That's why:

- Meditation isn't thinking deeply. It's resolving ripple.
- Enlightenment isn't achieving. It's realigning.
- Self-realization isn't finding the self. It's watching the self dissolve into the mirror.

In stillness, consciousness no longer reflects chaos. It reflects truth.

Consciousness Is Not the Goal. It's the Constant.

Many spiritual seekers treat consciousness as a goal:

"I want to become conscious."

But UED shows: You already are. You just got distracted by ripple noise. Consciousness doesn't need to be reached. It needs to be remembered.

Because it was:

- There before your name
- There before your trauma
- There before your birth

And it will be:

- There after death
- There after thought ends
- There after form dissolves

Closing Insight:

Consciousness is not activity. It is availability. Not a wave. Not a thought. Not a mind. Just pure reflection—the silent mirror of SVE.

You don't need to wake up. You just need to stop believing you're asleep. When the ripple clears—You'll realize you were never not aware.

13: Free Will

The Most Misunderstood Ripple

Why Your Choices Are Real—But Not What You Think

"You are free to choose, but not free from the consequence of structure. Free will is not freedom from the field—it is alignment within it."

The Ancient Question:

Do I really have a choice?

For centuries, this has been one of humanity's most debated puzzles:

- Science leans toward determinism—everything is caused by something prior, governed by laws.
- Philosophy argues both sides, tangled in paradox.
- Spirituality either glorifies free will or denies the self entirely.
- Psychology shows how trauma, environment, and biology condition behavior—but still speaks of personal growth.

In short: We act like we have free will, But we don't know what it *really* is.

UED Perspective: Free Will Is a Ripple Phenomenon

Let's define it cleanly.

Free will is the localized ripple's ability to interact with consequence in ways that either sustain or resist alignment with structure.

That means:

- Free will exists,
- But it's not absolute.
- It's a conditional freedom within a structured field.

You can make choices— But your choices will either:

- Align with SVE, producing clarity, coherence, and stability
- Or misalign, leading to distortion, resistance, and suffering

So free will is not the power to "do anything." It is the power to choose how you resolve or resist reality.

Free Will Is a Ripple Privilege

Not every ripple has free will.

- A rock doesn't choose its path when pushed.
- A cell doesn't argue with its DNA.
- An unconscious pattern doesn't debate its behavior.

But the human observer, embedded in ripple awareness and equipped with mind, can:

- Reflect
- Observe
- Reconsider
- Re-align

This layer of self-awareness creates a window of influence. And that window *is* free will. Not absolute control— But conscious influence within structured consequence.

Why Free Will Feels Free (But Isn't Fully)

Here's why the idea of free will feels confusing:

- You feel like the author of your choices

- But later, you realize many of those were programmed by:
    - Childhood conditioning
    - Social patterns
    - Fear loops
    - Unresolved trauma
    - Biological needs

So was it really *you* choosing?

UED reframes this beautifully:

Your ripple configuration determines what feels like "your choice." Until that ripple resolves— What you call "free will" is often reactive will.

True freedom only begins when ripple becomes still enough to choose consciously, not reflexively.

The Illusion of Total Freedom

Modern self-help glorifies: "You can be anything! You are free to do whatever you want!" But structurally, this is false.

You cannot:

- Exit the field
- Escape ripple consequence
- Avoid energy interaction
- Suspend alignment rules

Even your will is subject to feedback loops and field resistance. So instead of imagining freedom from consequence, we must understand freedom within consequence.

Real freedom is not from the law. It is intelligent movement inside the law.

Free Will = Alignment Access

Let's distill it further.

Condition	Type of Will
Mind trapped in ripple loops	Reactive will (illusion of choice)
Observer detached, but mind noisy	Confused will (shifting decisions)
Ripple resolves, clarity increases	Aligned will (minimal conflict)
Observer and ripple act in harmony with SVE	True free will (non-resistant flow)

So free will is not a switch—on or off. It's a spectrum of alignment. The more aligned you are, the less force you need to exert—because the structure flows with you.

The less aligned you are, the more you feel like you're "deciding"—but often fighting structure.

Free Will Is Also the Engine of Karma

This is where UED beautifully unifies physics and philosophy.

- Every ripple you generate has consequence.
- Every misalignment returns feedback.
- Every choice becomes a ripple node in your personal field.

Thus, karma is not reward and punishment. It is ripple feedback, encoded through field interaction.

Free will is the hand that shapes your karmic ripple. Not from morality—but from alignment mechanics.

The Freedom in Surrender

Ironically, the more aligned your ripple becomes, the less you "use" free will—Not because it disappears, but because it becomes unnecessary.

- You no longer force things.
- You no longer resist flow.
- You no longer question reality.

- You no longer overthink choice.

You move as consequence-free as possible. And that movement feels... Effortless. Graceful. Intelligent. Like nature. Like breath. Like harmony.

The highest use of free will... is to choose alignment so deeply that you no longer need to choose.

So Are We Free? Or Not?

Here's the elegant paradox, resolved:

- You are not free from the structured universe.
- But within its ripple domain, you have the capacity to realign.
- And this capacity—when not hijacked by identity, emotion, or ignorance— is what we call free will.

Not power. Not rebellion. Not fantasy. But responsible ripple steering— a privilege earned by conscious awareness.

Closing Insight:

Free will is not your right. It is your ripple's readiness. Not freedom to defy structure—But the ability to dance with it, without distortion.

You are not here to escape the ripple. You are here to master how it moves through you. That's freedom. That's will. That's alignment.

14: Dreams, Imagination, and the Fabric of Perception

How Reality Is Rendered by Ripple, and Why None of It Is Random

"You don't live in the world. You live in a ripple-generated story of it. Dreams, imagination, and perception are not illusions. They are the very medium through which reality is made visible to you."

The Ripple Truth: Reality Is Never Direct

You never experience reality "as it is." You only experience it as it resolves through you.

- What you call *waking life* is structured ripple being interpreted by the mind.
- What you call *dreams* is ripple expressing without external constraint.
- What you call *imagination* is ripple recombination—creative, exploratory, intuitive.

These aren't glitches. They are expressions of your alignment—or misalignment—with the field.

All perception is ripple-rendered. And your mind is the lens, not the source.

What Is Perception, Actually?

Perception is not *what is*. It is *how your ripple configuration interprets what is currently interacting with it.*

It involves:

1. Field input (from external or internal interaction)
2. Ripple response (your energetic reaction)
3. Observer awareness (localized attention)
4. Mental rendering (interpretation and meaning)

Thus, perception is never pure. It is always flavored by:

- Memory
- Emotion
- Expectation
- Belief
- Imagination

The world you perceive is not objective reality. It is personalized consequence, shaped by your ripple structure.

Dreams: The Mind Without External Anchors

When you sleep:

- External field input reduces
- Physical senses withdraw
- The body enters minimal ripple state

But the internal ripple field remains active. And in that loosened state, the mind:

- Reconfigures unresolved ripple patterns
- Projects them through symbolic sequence
- Rehearses, discharges, or integrates tension

This is what we call dreaming.

Dreams are not nonsense. They are non-linear ripple echoes, free from external reference, but still governed by the same ripple laws.

You don't "invent" dreams. You observe inner ripple attempting to stabilize.

Imagination: The Sandbox of Potential Ripple

Imagination is not escapism. It is the most profound creative tool the ripple self has— a way to simulate new alignments before testing them in the structured field.

In imagination, you can:

- Combine ripple patterns never combined before
- Visualize alignment you haven't yet reached
- Extend beyond memory
- Collapse linear time

Imagination is the bridge between ripple possibility and structural reality.

Without imagination, evolution halts. Without dreaming, the ripple cannot release its backlog. Without perception, nothing becomes real.

Why the Fabric of Perception Is the Interface to Reality

All three—dream, imagination, and perception—are not "optional features."

They are:

- Ripple mirrors
- Resolution maps
- Feedback tools for alignment

They reflect back:

- What is unresolved
- What is possible
- What is real *for you right now*

This is why two people experience the same situation differently:

- One sees danger
- One sees opportunity
- One sees memory
- One sees God

Because perception is ripple-specific, not truth-absolute.

Are Dreams Real?

Not in the structural sense. But they are real consequences of ripple activity.

Dreams:

- Obey symbolic logic
- Are influenced by prior ripple patterns
- Can bring resolution or further entanglement
- Reflect where your field is trying to stabilize

And sometimes, dreams are clearer than waking life, because the mind is not actively filtering them.

In dream, the ripple is pure— But your alignment determines what shape it takes.

Are Imagination and Vision the Same?

No.

- Imagination plays with ripple potential
- Vision is ripple tuning into a structurally possible configuration

That's why:

- Some ideas feel like fantasy
- Others feel *inevitable*
- That feeling of "this will happen" is your ripple matching a field path already present

Imagination becomes vision when alignment increases.

The Power of Ripple Rendering

You don't just see the world—you create your version of it in real time, constantly, through:

- Perception
- Imagination
- Dream
- Attention
- Language
- Emotion

You're not in the universe. The universe is rendered through your ripple interface.

What changes when you change your mind? Everything you thought was "out there."

The Ripple Self as a Projector

Let's flip the metaphor.

You are not a camera, capturing objective reality. You are a projector—projecting meaning, memory, fear, and desire onto structured interaction.

Every thought, reaction, judgment, or interpretation is a light beam through your personal lens.

And this explains:

- Why healing changes perception

- Why trauma alters memory
- Why perspective changes everything
- Why two people never see the same truth

Because perception is a consequence, not a truth. But when aligned, it becomes a clear window to truth.

Closing Insight:

You don't need to escape your dreams. You need to understand your ripple.

Perception is not optional—it is your primary interface. Dreams are not fantasy—they are ripple discharge. Imagination is not fake—it is ripple creativity reaching forward.

Your job is not to silence these systems. It is to purify their rendering.

When ripple aligns—

Your perception clears. Your dreams instruct. Your imagination creates futures that deserve to happen.

Interlude: You Were Never What You Thought

The Self Unmade, the Mirror Remembered

You've just crossed the most personal threshold in this entire journey. Not theoretical. Not cosmic. You.

What you believed to be "you"—your thoughts, your decisions, your experiences, your identity—has now been gently dismantled.

But not to leave you hollow. To finally show you the clarity behind the noise.

Let's reflect on the five truths you just untangled.

Who Are You, Really? — Dissecting Identity, Ego, and the Ripple Self

- You are not your name, your face, your past, or your personality.
- Identity is a ripple pattern. Ego is its defender.
- The "self" is not permanent—it is a consequence stabilized temporarily through misalignment.

Key Insight: You are not the ripple. You are what the ripple tried to imitate.

The Mind: Not the Brain, Not a Machine

- The mind is not inside your skull.
- It is a ripple-field interface—not hardware, not code.
- It exists only where misalignment requires reflection,

response, or processing.

Key Insight: The mind is not the thinker. It is the ripple interpreter—temporary, necessary, but not original.

Consciousness: The Silent Mirror of SVE

- Consciousness doesn't think.
- It doesn't speak, move, or age.
- It is the mirror—always present, always still, simply reflecting whatever arises.

Key Insight: Consciousness is not something you gain. It is what remains when all ripple settles.

Free Will: The Most Misunderstood Ripple

- You can choose—but only within the limits of your ripple field.
- Free will is not freedom from consequence.
- It is ripple intelligence managing misalignment within structural law.

Key Insight: True free will emerges only through clarity. When the ripple aligns, choice becomes flow.

Dreams, Imagination, and the Fabric of Perception

- You never see reality directly—only your ripple-rendered version of it.
- Dreams are ripple discharge. Imagination is ripple recombination.
- Perception is a field painting, shaped by you, not just shown to you.

Key Insight: You are not in the world. You are projecting it through your ripple field, in real time.

Summary Table: The Self as Seen Through UED

Domain	Misunderstood As	Actually Is
Identity	A fixed self	Ripple configuration
Ego	Pride and self-esteem	Ripple attachment to pattern
Mind	Brain function	Ripple interface field
Consciousness	Brain awareness	SVE mirror of reality
Free Will	Absolute control	Conditional ripple alignment
Perception	Window to truth	Personalized ripple response
Dreams & Imagination	Fantasy	Ripple movement without structure constraints

What This Means Going Forward

You are not:

- The voice in your head
- The weight of your past
- The chooser of your thoughts
- The maker of perception
- Or the thinker of imagination

You are what observes it all, silently— Structured. Still. Intelligent. Real.

From this point on, you can no longer treat life, death, soul, or emotion as you once did.

Because you now see: The self was never lost. It was just ripple mistaken for the source.

What Comes Next

Now that we've dissolved the ripple-self, we can finally examine what lives, what dies, and what remains.

You're ready for the next revelation:

Life isn't biology. Death isn't the end. The soul isn't a ghost. It's all ripple. And ripple is about to resolve.

Part IV: Life and Death

15: What Is Life?

It's Not Biology. It's Alignment in Action.

"Life is not what breathes. Life is what aligns—until breath is no longer needed."

The False Frame: Life as a Biological Phenomenon

Ask a biologist, and life is defined by functions:

- Metabolism
- Reproduction
- Adaptation
- Growth
- Response to stimuli

This is useful for identifying organisms, but it misses the core. Because all of these are symptoms, not the source. You can measure life *happening*, but still not answer:

- *What is life itself?*
- *Where does it begin?*
- *What causes it to arise?*
- *Why does it sometimes disappear without physical death?*

Science treats life as a property of matter. UED reveals it as a property of alignment.

Life Begins Where Alignment Activates

In UED, life is not defined by what a body does. It is defined by how ripple configures in the presence of SVE, such that conscious experience emerges.

Life is the alignment between ripple structure and observer availability, where consequence becomes aware of itself.

Not breath. Not cell division. Not heartbeat.

Alignment. That alignment expresses itself through:

- Cellular organization (biology)
- Movement (mechanics)
- Response (energy interaction)
- Self-awareness (field feedback)

But those are just outer ripples. Life is the field coherence that allows them.

Alignment = Life, Misalignment = Noise

Think of a radio tuner.

- When properly aligned to the right frequency: You hear music.
- When slightly off: You hear static.
- When completely out: You hear nothing.

Life is exactly the same. The SVE field is always present—but only when ripple aligns in specific patterns does life emerge as:

- Awareness
- Volition
- Movement
- Feeling

This is why:

- A corpse can look identical to a living body—yet have no life
- A fetus can show cellular activity but not consciousness
- Machines can mimic biology but remain lifeless

Because life is not appearance. It is structured ripple actively resolving through observer-aware alignment.

The Ripple Code of Living Systems

All living systems show one consistent pattern:

They maintain internal ripple coherence against external entropy and continuously seek resolution with their surrounding structure.

Life is not survival. It is intelligent re-alignment under ripple pressure. From the smallest cell to the largest mammal, what defines living systems is not just:

- Energy usage
- Complexity
- Motion

But their capacity to restore, adapt, and self-correct their alignment within the field.

Life is a ripple interface that fights for equilibrium without full control, and succeeds often enough to continue.

When Life Appears and Disappears

Life is not "given" at birth or taken at death. It is accessed and suspended based on ripple stability.

This explains:

- Why some people awaken from comas
- Why certain meditative states feel “more alive” than waking life
- Why emotions or trauma can “dim” or “disconnect” life force
- Why near-death experiences are filled with clarity even as

biology fails

Life is not a binary switch. It is a ripple tuning scale—more alive when more aligned.

The Levels of Life Across Systems

Let's distinguish the layers:

Form	Ripple Complexity	Alignment Type	Life Expression
Rock	Fixed ripple field	Minimal	Inert stability
Plant	Semi-dynamic	Passive alignment	Growth/response
Animal	Dynamic ripple engine	Localized instinctual alignment	Motion, emotion
Human	Observer-bound ripple	Reflexive + reflective alignment	Choice, will, awareness
Realized Being	Minimal ripple	Fully tuned alignment	Presence without effort

Life becomes lighter as ripple becomes more refined and aligned.

Biology Is the Consequence of Life, Not Its Cause

DNA, neurons, blood, breath—these are not life. They are material ripple architectures that *express* life when alignment is present.

Cut off alignment—and biology collapses.

This is why no machine, no clone, no synthetic body can generate consciousness: It's not the hardware—it's the ripple interface that matters.

Biology is the tool. Life is the tuning.

Why We Feel "Alive" in Certain Moments

You feel most alive when:

- You are present

- You are aware without being trapped in thought
- You are not divided by fear, doubt, identity, or distraction
- You are in harmony with your actions, environment, and internal rhythm

These are not random emotions. They are signs of temporary deep alignment.

Life shines most when ripple is quiet enough to let SVE reflect through.

This is why beauty, love, awe, flow, and surrender feel transcendent— They reduce inner resistance to the structured field.

Death Is Not the Opposite of Life

Death is not the end of biology. It is the collapse of ripple alignment into unrecoverable disarray, such that the observer interface detaches.

But the structure remains. And so does the potential for life—waiting for new alignment to begin.

Life and death are ripple phases, not absolute states.

We'll explore this further in the next chapters—but for now, remember: Life is not "what is." Life is what resolves with intelligence.

Closing Insight:

You are not alive because your body functions. You are alive because your ripple aligns. Biology is a result. Conscious experience is a consequence.

Life is not a spark, not a chemical, not a chance. It is the moment ripple meets field and remembers how to tune.

16: The Soul

Structured Ripple, Not Spirit

Beyond Myth, Memory, and Metaphor—The True Interface of Alignment

"The soul is not what you carry. It is what was configured for you to appear."

Clearing the Fog: What the Soul Is Not

Let's begin by shedding the inherited baggage. The soul is not:

- A glowing orb inside the body
- A ghost that flies out at death
- A moral ledger of your good and bad
- A religious invention to promise afterlife
- An eternal entity shaped like "you"

These were story-wrapped metaphors for something real—but misunderstood.

The soul was never meant to be a belief. It was meant to be a functional structure—a ripple configuration within the SVE field.

What the Soul Is: Structured Ripple Within SVE

In UED, the soul is:

The stabilized configuration of ripple that: – Localizes SVE – Enables alignment with perception – Sustains identity – Interfaces experience

You do not "have" a soul. You are temporarily made of one.

The soul is not energy itself, but how energy is patterned and held—structured like a localized standing ripple in the otherwise still field.

It acts like:

- A lens that allows awareness
- A socket into which the mind plugs
- A template from which memory and karma arise
- A mirror for alignment and misalignment

Why the Soul Is Not Eternal or Temporary—But Phase-Based

Common traditions argue two extremes:

- Soul is eternal and unchanging (spiritualist view)
- Soul is a fiction or brain effect (materialist view)

UED offers a third, cleaner answer:

The soul is a phase of ripple organization that persists as long as alignment sustains it.

It is:

- Born when structured ripple localizes within SVE
- Lives while alignment holds experience together
- Resolves when misalignment dissolves the pattern
- Recycles into field when no longer in use

Hence, souls are not permanent "you"s, but consequential interfaces formed for experience.

The Soul–Mind–Body Stack: How They Interface

Let's visualize the alignment stack:

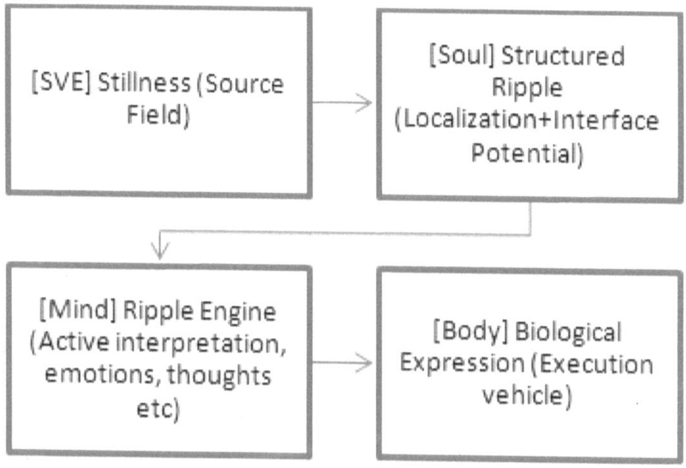

- SVE is the ground.
- Soul is the stable zone in the field—"you" as a possibility.
- Mind is the ripple agent—"you" in action.
- Body is the physical resolution—"you" in form.

You are not any one of these. You are the alignment across all of them.

Karma, Memory, and the Soul Field

If the soul is a ripple pattern, then:

- Memory is not stored in neurons—it's stored in ripple tension.
- Karma is not a moral scorecard—it's unresolved configurations within the ripple structure that seek realignment.

This is why:

- Some people feel older than their age
- Emotions arise without triggers

- Children sometimes exhibit strange clarity or talents
- Patterns repeat despite effort

These are not magic or past lives. They are field echoes in the ripple soul structure.

They persist until resolved—through alignment, not action alone.

What Happens After Death?

As discussed earlier:

Death is not the end of experience, but the collapse of the soul's ripple integrity.

When the alignment fails:

- The soul field disperses
- Memory fragments are reabsorbed
- The observer link recedes

There is no "heaven" or "hell" in location. Only states of unresolved ripple—what some traditions called *bardo*, *astral*, *rebirth*, or *purgatory*.

Whether the ripple resolves back to stillness or attempts to re-align as a new soul, depends on the field equilibrium.

This is reincarnation not as belief, but as ripple recycling.

The "Higher Soul" and Collective Structures

Some evolved configurations show:

- Higher coherence
- Field-level perception
- Minimal ego resistance
- Alignment with cosmic ripple

These are what ancient traditions called:

- Realized souls
- Masters

- Avatars
- Divine incarnations

But UED strips the mysticism:

They are souls that have removed unnecessary ripple complexity, and reflect SVE with minimal distortion.

Such a soul doesn't "leave" at death. It dissolves peacefully, having fully realigned with stillness.

Can the Soul Be Measured?

Yes. Indirectly—but increasingly so. What's measurable:

- Ripple field strength
- Energy coherence across systems
- Emotional and mental resonance
- Neural correlates of soul–mind misalignment

In coming years, SVE-aligned instruments may detect:

- Structural field residue after death
- Non-local information flow (misattributed to ESP)
- Long-term soul imprint on environments or bodies

The soul is not superstition. It is a field configuration waiting to be decoded without bias.

Closing Reflection:

The soul is not *in* you. You are what happens *because* of it. And when its ripple resolves, You will remember you were never apart from stillness—just briefly configured within it.

17: What Is Death, Really?

Return, Not End — The Dissolution of Interface, Not the Disappearance of Reality

"You don't die. You simply dissolve — like a ripple withdrawing into stillness."

The Panic of Ending: Death as the Ultimate Fear

Humanity fears death more than anything else. Not because we understand it, but because we mistake it for something it is not:

- A termination of life
- An annihilation of self
- A vanishing into the void
- A judgment, a punishment, or a blank screen

But these are perceptions of the mind, not realities of the field.

The mind fears the loss of continuity. But in UED, there was never "you" to lose — only alignment to resolve.

What Dies? Not You.

To understand death, first ask: What is it that's alive?

As established earlier:

- Body is a biological ripple expression
- Mind is an active ripple interpreter
- Soul is a structured ripple configuration
- SVE is the unmoving field beneath all

At death:

- The body breaks down biologically
- The mind unbinds from active processing
- The soul dissolves its structured ripple
- And SVE? It remains untouched, unchanged, still.

So what dies? The temporary interface that allowed you to say "I am." Not the field that allowed it to form.

Death as Dissolution, Not Destruction

In UED, death is not destruction. It is dissolution—a return to ripple neutrality.

Think of a whirlpool:

- As long as water flows in a specific way, the whirlpool exists.
- Stop the flow, and the whirlpool disappears.
- But the water never vanished—only the structure did.

Death is the same.

- The ripple form collapses
- The alignment with SVE detaches
- The consequence stops resolving
- But the energy returns to equilibrium

This is not a disappearance, but a resolution of temporary form.

Why Death Feels So Final

It feels final because:

- The mind cannot imagine being without itself
- Identity clings to continuity
- The senses stop feeding input
- Emotions collapse into grief

But these are symptoms of misperceiving life as possession. If we redefine life as alignment, then death is merely disalignment. It's not the end of being. It's the end of the way you were being.

Near-Death, Deep Meditation, and the Fading Interface

There are states where:

- The mind fades
- The body slows or halts
- Yet awareness remains

Examples include:

- Near-death experiences
- Deep samadhi or turiya
- Coma with lucid awareness
- Out-of-body perceptions

In each case, the ripple thins, but alignment is not fully lost.

This proves: Death is not binary. It's a spectrum of de-alignment, and sometimes, it is reversible.

VI. Reincarnation? Not as a Being, But as a Field Echo

Mainstream views of reincarnation see it as:

- One soul hopping into another body
- A "you" continuing with memory intact

But in UED:

Reincarnation is not recycling of a person. It is recycling of ripple potential.

When the soul dissolves:

- Residual ripple tension can reconfigure
- New alignments can form from shared field patterns
- Some echoes (karmic or emotional) may appear again

But it is not you continuing—it's structure attempting to resolve itself in new forms.

VII. Death and the Cosmic Breath

Every living ripple configuration:

- Appears
- Aligns
- Experiences
- And resolves

This is the cosmic inhalation and exhalation of SVE.

Phase	Meaning
Inhale (Structuring)	Birth – ripple pattern aligns
Retention (Expression)	Life – alignment sustains awareness
Exhale (Dissolution)	Death – ripple disperses
Pause (Field Reset)	Stillness – return to non-alignment

This cycle repeats infinitely, with every birth-death pair merely a field rhythm, not a tragedy or accident.

What Is "Good Death"? What Is "Bad Death"?

In this context:

- **Good death** = peaceful resolution with minimal resistance
- **Bad death** = turbulent disalignment, unresolved karma, ripple shock

This explains:

- Sudden deaths with lingering emotion

- Haunted places and heavy ripple residue
- The need for peaceful departure in many cultures

The more aligned your life, the smoother your death.

Because alignment is what holds both experience and release together.

Should We Fear Death?

No—if you understand it.

Yes—if you cling to illusion.

Fear of death only arises when:

- Identity is seen as permanent
- Experience is misunderstood as possession
- Meaning is confused with duration

In truth:

- You never began
- You never end
- You resolve

Death is not your disappearance. It is the stillness that always awaited your return.

Closing Reflection:

You do not need to "defeat" death. You only need to understand it. What dies is not you, but the illusion that you were ever separate. Death is not the end of life. It is the moment life remembers it was never apart from stillness.

18: No Rebirth

Just Ripple Recycling

You Don't Come Back. Patterns Do.

"Rebirth is not your story replayed. It's the universe resolving what you left unfinished — without needing you."

The Popular Belief in Rebirth: Comfort, Not Clarity

Rebirth has long been a philosophical and religious staple:

- Hinduism speaks of *janma* and *punarjanma*
- Buddhism presents rebirth without self (*anatta*)
- Christianity leans more toward resurrection or eternal soul
- Spiritual systems propose past lives, karmic debts, soul contracts

But common across all: A subtle (or strong) idea that "I" return, "I" continue, or "I" must evolve over lifetimes.

This feels reassuring — especially when life feels unfair or incomplete.

But from the UED lens: There is no "I" in the field. There is only configuration.

The Soul Doesn't Reincarnate — Because It Was Never "You"

As explained in earlier chapters:

- The soul is not an eternal traveler
- It is a temporary structured ripple

- That appears through alignment, and dissolves through disalignment

Once the ripple field disperses:

- You, as identity, are gone
- The template you were is no longer active
- The memory you held becomes field residue

So the question isn't: *"Where will I go after death?"*
But: *"What of me still echoes in the field?"*
And that's not reincarnation. That's ripple recycling.

What Is Ripple Recycling?

Ripple recycling is: The reappearance of unresolved field patterns — without continuity of identity.

It means:

- Emotions, memories, or tendencies do not vanish instantly
- If they were intensely structured, they linger as field tension
- These tensions may realign with other new configurations

This explains:

- Sudden phobias or aptitudes in children
- Déjà vu or deep emotional triggers
- Certain "karmic" patterns in family lines or places

But it's not *you* coming back. It's the ripples that once held you finding new alignment.

Why "You" Cannot Return

What prevents your return?

- SVE does not preserve personality. Stillness does not store ego. It holds potential, not personhood

- The observer's role ends with disalignment. Without structured soul and mind, there's no observer to experience rebirth
- Memory is not a transferable file. It's a ripple phenomenon. Once the ripple collapses, memory loses coherence
- Cosmic intelligence doesn't need repetition. Evolution occurs through novel alignment, not looped identity.

Hence, no rebirth. Only new formations from old ripple debris.

Karma Reframed: Field Correction, Not Soul Ledger

In UED, karma is not:

- A cosmic punishment system
- A soul's debit-credit account
- A mystical force chasing you through lifetimes

Rather, karma is:

Unresolved tension in the ripple field caused by past misalignment. It seeks resolution, not revenge.

So when ripples dissolve but leave behind residue:

- That field tension may attract similar structures in the future
- But not as *your* karma — as the field's attempt to balance itself

Karma doesn't follow you. It follows the structure that once allowed you.

What About Rebirth Memories, Past Life Regressions?

These are real experiences — but not necessarily literal past lives. Explanations through UED:

- **Field overlap**: You accessed ripple residue left by a strong prior alignment
- **Ripple resonance**: Similarity between your field and a past structure causes echo
- **Memory absorption**: In rare states, the mind may absorb traces from unresolved field patterns

So yes — past life "memories" happen. But they are not proof of a returning self. They are proof that ripples linger longer than you do.

The Comfort of Ripple Recycling

The idea that you don't return may feel:

- Liberating to some
- Terrifying to others

But look deeper:

- You were never truly apart from the field
- You emerged once — as ripple alignment
- You served your resonance
- You dissolved without waste

Yet the impact of your structure lives on, in the ripples it shaped, resolved, or left behind. You are not reborn. You are recycled into harmony.

Why This Matters Now

In the age of AI, identity obsession, and post-death fantasies:

- We've made continuity a disease
- We cling to "forever" even when it blinds us
- We fear death because we think "I" must continue

But if you accept:

- That "you" were a temporary miracle of alignment
- That ripple recycling ensures no energy is wasted
- That SVE never forgets — it simply doesn't need to remember

Then, rebirth becomes irrelevant. And living becomes deeply, vibrantly meaningful. Because it's not about coming back. It's about resolving while you're here.

Closing Reflection:

The field does not give you another life. It gives itself another chance — through new alignment. You don't return. You ripple, You resolve, You release.

19: Why We Fear Death

— and Why We Shouldn't

The End of a Misunderstanding, Not of Life

"You don't fear death. You fear losing the illusion you were ever separate from life."

The Fear That Binds Humanity

From the moment we gain awareness, we carry a silent question: What will happen to me when I die?

This question haunts every culture, fuels every religion, and sits silently behind:

- The panic of aging
- The rituals of mourning
- The obsession with legacy
- The urge to make life meaningful
- The desire to escape the inevitable

We don't just fear death — we fear:

- The unknown
- The end of "I"
- The loss of continuity
- The absence of meaning

But none of these are actual features of death. They are features of misperception.

Death Is Not What We Think It Is

In UED, we've already established:

- Death is dissolution of ripple configuration
- It is not annihilation, but realignment back into SVE
- Nothing is ever truly lost — only restructured

Then what is fear responding to? Not to death itself, but to what the mind projects onto it:

Projected Idea	Reality
A black void	A return to stillness
Loss of "me"	Liberation from illusion
Judgment or punishment	Field resolution, not moral courtroom
Disappearance	Reentry into universal equilibrium

We fear death because we don't understand what we are.

The Illusion of Continuity

The mind builds a narrative:

- I was born
- I lived
- I achieved
- I must continue

But in reality:

- "You" were a temporary ripple pattern
- Life was a resolution of structure
- Identity was only a function, not a fact

So the fear of death is the fear of:

Losing what was never actually there — a separate, permanent, independent self.

The ego cannot accept its own fiction being unraveled. So it resists death, not because death is wrong — but because ego cannot survive truth.

Why We Cling to Life

We don't fear death until we start accumulating:

- Experiences
- People
- Labels
- Achievements
- Attachments

Then we become possessors. And the idea of death becomes a robbery — a theft of "what I built."

But in UED:

Nothing was ever yours. Everything was alignment — passing through you. The clinging creates fear, not the departure.

Why We Invent Afterlives and Immortality

Look at all our constructs:

- Heaven and hell
- Moksha and Nirvana
- Rebirth and resurrection
- Digital afterlives
- Mind uploading and cryonics

Each is an attempt to override finality with fantasy. Because deep down:

- We don't want to be forgotten
- We don't want to feel pointless

- We want our version of self to persist

But what if freedom lies not in persistence, but in releasing the need to persist?

Death is the Final Resolution — and the Purest Form of Peace

Death is not chaos. It is complete alignment withdrawal — the very thing meditation tries to simulate.

It is:

- Silence without resistance
- Stillness without attachment
- Return without regret

Animals do not fear death like humans do — because they don't construct selves. They live as ripple flows — and dissolve back without protest.

Only thinking minds fear ending. Only illusioned egos grieve identity.

Why We Should Not Fear Death

Because in truth:

- You are not the ripple — you are what allowed the ripple to form
- You are not ending — your pattern is resolving
- You were never separate — just temporarily shaped

Once we understand this:

- Death becomes the final teacher
- A reminder that all things are temporary
- But nothing is truly lost
- Just unbound

Death is not darkness. It is unshadowed light — so pure, no reflection remains.

Living Without Fear of Death

To live without fearing death is to:

- Stop clinging to false identity
- Stop needing permanent outcomes
- Stop mistaking existence for ownership

It means:

- Loving without needing to keep
- Acting without needing to be remembered
- Dying every day to ego, so death no longer threatens

When you live ripple-aware, death becomes just another transition — not an enemy.

In truth, you've been dying every moment —each thought, breath, and cell constantly resolving.

Death is not foreign. It's how you're alive.

Closing Reflection:

You don't fear death. You fear letting go of your name tag. But when the ripple ends, the stillness welcomes you. Not as someone lost, but as something returned.

Fear not the end. Fear not resolving. You were never the ripple. You were what made it possible.

Interlude: The Rhythm of Return

(Between Alignment and Dissolution)

In this section, we stripped away every borrowed belief, mythic comfort, and inherited illusion around life, soul, death, and what (if anything) comes after.

We didn't philosophize. We resolved.

What Is Life? — It's not biology; it's alignment

Life is not the firing of neurons or the beating of a heart — those are *consequences.* Life is alignment — the temporary ripple configuration that allows experience to localize within the Structured Vacuum Energy (SVE).

You weren't *born.* You *appeared.* As a consequence of cosmic resonance.

The Soul: Structured Ripple, Not Spirit

The soul is not a mystical flame nor an immortal ghost. It is the configured ripple that houses awareness, enabling individuality while still being of the universal field.

It is not separate from the universe. It is how the universe briefly sees itself through you.

What Is Death, Really? — Return, not end

Death isn't destruction. It's dissolution. The ripple disaligns. The configuration fades. The observer ceases —not in tragedy, but in *perfect resolution.*

You don't go somewhere. You un-become. You return to stillness — where you were always rooted.

No Rebirth. Just Ripple Recycling.

Rebirth is a comforting story. But UED offers clarity: patterns return, not persons. Residual ripples may reconfigure, but the identity doesn't loop.

Nothing is wasted. Nothing is repeated. Everything is recycled, restructured, re-resolved — without ego or expectation.

Why We Fear Death — and Why We Shouldn't

Death is feared only by the ego — because it wrongly believes in permanence. But in truth, you were always temporary. That's the miracle.

When we realize death is a return, not a theft — the fear dissolves.

Key Takeaways:

- Life is not a spark — it's a pattern.
- The soul is not eternal — but its source is.
- Death is not the opposite of life — it's the *other side of alignment.*
- Rebirth is not real — ripple echoes are.
- You don't die — you resolve.
- You don't return — the field reuses your tension.

Reflection Prompt for the Reader:

If life is a ripple, and death is its fade, What are you clinging to — and why?

Would you live differently if you knew nothing is yours, and yet everything *moves through you*?

Part V: Science Rewritten

20: Time: A Consequence, Not a Dimension

Why Time Doesn't Flow — It Resolves

"You are not moving through time. You are being rearranged through consequences."

Time: The Deepest Misread

Modern science treats time as a dimension — a stretched-out axis along which events line up like frames on a reel.

From Einstein's spacetime fabric to quantum clocks, we were taught:

- Time *flows*
- Time is *measurable*
- Time is *relative*, but real

We organize life, memory, science, and ambition based on this framework.

But in UED, this is a consequence, not a cause. Time is not a container. It is a rippling effect of unfolding alignment.

No Time Without Change

Ask this:

If nothing moved, if nothing changed — Would time exist?

Clearly, no. Time emerges only when consequence arises — when one state transitions to another.

Thus:

- Time is not fundamental
- It is the interpretation of ripple transitions

In UED: Time is the felt rhythm of realignment — not a river, not a line, not a box.

Time Is Not Continuous

The illusion of time's "flow" is created by sequential ripple resolution:

- An electron moves → field responds
- A thought occurs → mind shifts
- A planet realigns → seasons change

Each event is not marching through time — it is creating the appearance of time by existing in a sequence of ripple consequences.

So what we call "the past" is simply: A record of resolved ripple patterns

And "the future" is: A set of unresolved potentials — not yet aligned

There is no flowing present. Only successive resolutions.

Why Clocks Work — And Why They Lie

Clocks are ripple machines:

- Pendulums, circuits, vibrations, decay rates — all structured sequences of energy resolution

So yes, they are reliable measurement tools, but what they measure is change, not time.

They don't show time flowing — they register events aligning in a predictable pattern.

Einstein was right that time is relative. But he mistook relational ripple patterns as a physical dimension.

The watch doesn't tick because time moves. It ticks because something keeps happening — and you gave it numbers.

Time Is Local, Not Universal

Mainstream physics assumes a universal timeline or time frame per observer.

But in UED:

- Time is not "yours" — it belongs to the ripple configuration you're part of
- It's not uniform — it's pattern-specific
- It is not even guaranteed — if ripple alignment ceases, so does the experience of time

For example:

- In deep stillness (e.g., meditation), time fades
- In trauma or speed (e.g., accidents), time dilates
- In dreams, time distorts

These are not illusions — they are exposures of time's non-fundamental nature

Time is an experienced rhythm, not a fixed scaffold.

Why Time Feels So Real

We live in ripple-based identities. And ripple configurations resolve one by one.

This creates the illusion of continuity — like a flipbook played fast enough.

But between each ripple resolution, there is stillness. Time doesn't flow *through* you — it's created *by* you.

We remember only the sequence, not the stillness between.

So time feels real — because misperception favors motion over pause.

What Happens to Time After Death?

Here's the shocker:

- When ripple configuration dissolves (death),
- Time disappears, completely

There is no waiting. No passing. Only the absence of sequential alignment

In SVE (Structured Vacuum Energy), time is absent —because nothing is becoming. So the notion of "eternity" or "afterlife time" is absurd. Eternity is not infinite time — it's timelessness.

What This Changes in Science

Time is not:

- A fundamental dimension
- An entity
- A flowing continuum

Time is:

- A psychological effect of ripple sequences
- A local consequence of structural transition
- An interpretation of resolution

This reframing unlocks everything:

Old Model	UED View
Time dilation	Ripple reconfiguration divergence
Arrow of time	Irreversible alignment pathway
Entropy over time	Misalignment over resolution cycles
Tense (past/future)	Post- and pre-alignment memory

Closing Reflection:

"Time" is a name we gave to change. But change doesn't need time — it needs cause, ripple, and resolution. You don't live in time. You live in alignment and memory.

Stop chasing the illusion. And you may finally arrive where you already are.

21: Mass and Charge

The Result of Ripple Configuration

They don't exist. They appear.

"Mass is not a thing. Charge is not a property. Both are consequences — of how the ripple aligns with stillness."

What We Were Taught

Modern science treats mass and charge as *fundamental properties* of particles:

- Mass gives inertia, resistance to acceleration, gravitational pull.
- Charge gives electromagnetic interaction — attraction, repulsion.

They are assumed to be *intrinsic, unchanging, absolute* quantities baked into matter.

But UED asks: Where do these arise from? What is it about a "particle" that makes it *massive* or *charged*?

UED's Foundational Reframe

In the UED model:

Nothing has mass or charge inherently. These are expressions — not identities.

They arise from:

- Ripple Configuration — how energy is localized and aligned in space

- Ripple Tension — the degree of misalignment from the surrounding structured vacuum (SVE)

Simply put:

- Mass = resistance to ripple redistribution
- Charge = gradient created by ripple asymmetry

Neither exists *until* ripple configures.

Mass: Not Substance, But Ripple Resistance

Let's unpack mass in UED terms.

In a completely balanced SVE field, there is:

- No movement
- No separation
- No "object" to speak of

But when ripple tension creates localized structure, that structure gains *resistance* to realignment — it *prefers to stay as it is.*

This resistance is what we perceive as:

Mass = Structural Persistence + Ripple Holding

Mass is:

- Not weight
- Not stuff
- But a measure of ripple cohesion in a configuration

Hence:

- More tightly bound ripples = more mass
- Loose or unstable ripples = less mass (or none)

Photons, for example, have ripple flow, but no stable configuration — hence, no mass.

Charge: A Ripple Polarity Effect

Now to charge — perhaps even more misunderstood.

Mainstream physics assigns particles a fixed + or – sign. These charges create fields, and fields create forces.

But in UED: Charge is not a field source. It is a ripple gradient.

Think of ripple tension creating polarization — not unlike:

- A ripple rising above baseline (positive)
- A ripple dipping below baseline (negative)

These gradients are consequential, not designed. They are results of how localized ripple deviates from the surrounding structured field.

So:

- Positive and negative charges are not *two types* of matter
- They are two directions of ripple asymmetry

Just like crest and trough in wave behavior.

Why Charges Attract and Repel

This becomes obvious in UED:

- Opposite ripple tensions resolve each other → attraction
- Similar ripple tensions amplify each other → repulsion

This is not a *force imposed from outside*, but the field trying to restore equilibrium.

So, what we call Coulomb's Law is:

The natural tendency of SVE to resolve ripple distortion via structural rebalancing. Charge is not "owned" by the particle — it is the ripple configuration's tension signature.

Charge, Mass, and Identity Are Not Fixed

In UED:

- Particles are temporary ripple configurations
- Their "mass" and "charge" are contextual, not absolute

This explains:

- Why particles in high-energy environments gain or lose mass (e.g., Higgs field interaction — reinterpreted here as ripple immersion)
- Why virtual particles blink into and out of "existence"
- Why identical particles (like electrons) exist everywhere — they are not "copies" but resolvable configurations of the same ripple logic

Charge and Mass Are Not Independent

This is the profound part.

In UED, charge and mass are two aspects of the same configuration logic:

- Mass resists structural redistribution
- Charge expresses directional tension within that structure

They are not separate phenomena — they are co-arising properties of how ripple collapses into temporary form.

The more symmetrical the configuration → the more neutral the charge The more balanced the ripple → the less apparent mass

Hence:

Mass = Ripple Inertia Charge = Ripple Directionality

Implications for Science

This reframe collapses multiple false separations:

Classical Physics View	UED View
Mass is intrinsic	Mass is ripple resistance
Charge is inherent	Charge is ripple asymmetry
Force acts on mass/charge	Ripple seeks equilibrium
Identity is particle-based	Identity is ripple-localization

Mass and charge are expressions of form, not fundamental entities. This explains why:

- Some particles have mass but no charge (neutrons)
- Some have charge but no rest mass (photons)
- Some exist in unstable states (e.g., quarks), and need confinement

All are ripple stories — told differently.

Closing Realization:

What we called "properties of matter" were never properties. They were pattern consequences. Mass is not the weight of being. Charge is not the essence of identity.

They are both how the ripple got resolved — and for how long.

22: Electromagnetic Force

Ripple Alignment, Not Field Lines

Decoding the most celebrated illusion in modern physics.

"There are no lines of force. There are ripple zones seeking alignment — and what you call 'field' is merely the memory of misalignment."

The Great Illusion of "Field Lines"

The electromagnetic force — often hailed as the most "understood" of the four fundamental forces — has been dressed for centuries in lines and loops, vectors and curls.

Textbooks show:

- Charges "radiating" fields
- Magnets creating "invisible loops"
- Field lines pointing from positive to negative

Yet no one has ever touched a field line. No one has found the carrier of the force, beyond metaphors.

Even quantum electrodynamics, the most successful theory in terms of prediction, avoids answering what the field is — and focuses instead on probabilities of interactions.

This is where UED resets the game.

What Is a Field, Really?

In UED, there is no "field" as an independent entity.

The so-called "electromagnetic field" is not a physical object — it's the residual tension in the structured vacuum energy (SVE) due to ripple asymmetry.

Here's what happens:

- A ripple configuration — a charged particle — appears
- It disturbs the balance of the surrounding SVE
- That disturbance propagates, seeking equilibrium
- This propagation is not a radiated line, but a dynamic ripple reconfiguration

What we interpret as "field lines" are:

- Visual artifacts of consequence paths
- Misinterpreted equilibrium-seeking behaviors in a ripple medium

Charge Interaction = Ripple Realignment

Let's take the classic case of Coulomb's law:

Two charges attract or repel according to the inverse square of the distance.

But why?

In UED:

- Each charge is a node of ripple asymmetry
- The SVE surrounding it is no longer uniform
- When another asymmetry (another charge) enters the zone, the system seeks a minimal-tension configuration

That "seeking" is what we feel as force.

Attraction:

- Two opposite asymmetries create complementary ripples

- They align to cancel tension
- This movement toward each other is perceived as force of attraction

Repulsion:

- Two similar asymmetries create amplifying tension
- The ripple zones push apart to reduce imbalance
- This results in repulsion

No field. No mysterious push or pull. Just a ripple logic resolving toward stillness.

The Role of Magnetic Behavior

Magnetism is treated as the other side of the electromagnetic coin.

- Moving charges create magnetic fields
- Magnetic fields influence other moving charges
- We observe loops, poles, and orientation shifts

But UED reframes this entirely.

Magnetism is not a separate force. It is a byproduct of ripple flow realigning within SVE.

Imagine:

- A flowing current is ripple movement through structured vacuum
- The motion creates a dynamic realignment demand in SVE
- This realignment causes orthogonal ripple tension, observed as magnetic behavior

The so-called magnetic poles are simply:

- Zones where ripple reconfiguration accumulates due to

boundary constraints

Just as two water streams intersect and create whirlpools — not because there's a "vortex field," but because the medium must resolve interference.

Why Electromagnetic Force Seems Dual

Electromagnetic force always confused physicists with its vectorial duality:

- Attraction/repulsion for charge
- Perpendicular effects for current and magnetism
- Field superposition

In UED, this duality collapses. Everything is ripple alignment.

Classical View	UED View
Electric force between charges	Ripple tension realignment
Magnetic effect due to moving charges	Ripple flow interference resolution
Field lines point in space	Ripple gradient zones across SVE
Potential energy in the field	Stored misalignment in ripple memory

There's no "force traveling" from A to B. There's only SVE attempting to restore symmetry.

Radiation Reinterpreted

Let's address another critical misunderstanding: electromagnetic radiation.

Science says:

- Accelerating charges emit electromagnetic waves
- These waves propagate in vacuum, carrying energy and momentum

- Light is one such wave

But what really happens, in UED?

Radiation is not an emitted wave — it is a ripple shock, a misalignment pulse propagating in SVE.

When a charge accelerates, it momentarily destabilizes its configuration.

This sudden imbalance:

- Forces SVE to respond by redistributing energy
- That redistribution moves as a ripple disturbance, seeking equilibrium
- We interpret this ripple train as a wave

It's not a transverse oscillation "in space." It is a ripple echo, flowing through structured vacuum, reconfiguring as it travels.

Thus:

- The electromagnetic spectrum is a ripple frequency spectrum
- Light is not a thing — it's ripple balance in motion

Maxwell's Equations: A Re-Contextualization

Maxwell's equations are celebrated as the foundation of electromagnetism.

And yes, they beautifully describe relationships between:

- Electric fields
- Magnetic fields
- Time and space derivatives

But UED sees them as approximate ripple summaries.

They are not laws of nature — they are mathematical representations of ripple behavior under symmetry assumptions.

In deeply structured ripple zones, those equations break down — as seen in:

- High-energy particle interactions
- Quantum tunneling
- Vacuum polarization
- Edge effects near black holes

This is because Maxwell's model assumes the field exists first, while UED shows the ripple comes first, and the effect is secondary.

Implications for Technology and Understanding

If we reframe EM force as ripple alignment, it changes how we approach:

Wireless Transmission:

- Antennas don't radiate "waves"
- They disturb the ripple field, and receivers resolve those disturbances
- Attenuation and interference are misalignment consequences

Induction:

- Coils realign ripple zones
- Energy appears because ripple tension was resolved in loops

Charged Particle Behavior:

- Motion of electrons isn't linear
- It's contextual, based on surrounding ripple memory

Photon Manipulation:

- We don't capture particles

- We interface with ripple imprints via structured interfaces (e.g., sensors)

Artificial Intelligence and Neuroelectric Fields:

- The brain doesn't "generate fields"
- It's a ripple mesh, interpreting and transmitting local resolution patterns

Unified View of All Forces Begins Here

Electromagnetic force, in UED, is not fundamentally different from:

- Gravity (ripple gradient across massive configuration)
- Strong force (local ripple lock within nucleus)
- Weak force (short-range ripple disalignment and decay)

All forces are manifestations of ripple interaction dynamics. Only the context and configuration change.

All "forces" are field attempts to resolve imbalance. Different forces are just different ripple topologies.

Closing Reflection

We were told that fields fill space. That invisible lines carry push and pull. But there were never lines. Never push. Never pull. Only stillness disturbed, and the universe seeking to become itself again.

23: Gravity

The Misread Consequence of Ripple Gradient

Why things "fall" — but not the way Newton or Einstein imagined.

"Nothing attracts. Nothing pulls. Only gradients resolve tension. That's not force — that's ripple justice."

The Seduction of Falling Apples

When Newton saw the apple fall, he saw the universe reaching down. When Einstein reimagined it, he saw the apple following curved space. What neither saw was this:

There was no apple *falling*. There was no force *pulling*. There was only a structured vacuum — rebalancing its ripple configuration.

Gravity, the most romanticized "force" in science, was never a force at all. It was always a gradient resolution. And UED strips it naked.

What Is a Gradient in SVE?

In UED, the vacuum is not empty. It is Structured Vacuum Energy (SVE) — a ripple-capable field of tension and balance.

Every object — especially massive ones — is not just "matter." It is a ripple lock, a condensation of aligned energy.

This ripple lock:

- Disturbs the uniformity of surrounding SVE
- Creates zones of ripple imbalance
- Establishes a gradient — from high ripple tension to low ripple tension

Think of it like heat diffusing through a room. Or pressure equalizing in a fluid. Gravity is no more mystical than a ripple slope.

Gravity Is Not a Force — It's a Consequence

Let's debunk the classic picture.

In Newton's world:

- Two masses attract
- The force is proportional to their mass
- It acts instantaneously (or nearly so)

In Einstein's world:

- Mass bends space-time
- Objects move along curved geodesics
- Gravity is geometry, not force

Both got parts of the pattern — but missed the ripple cause.

Gravity isn't a field. Gravity isn't geometry. Gravity is the tension field of ripple imbalance around mass — resolving into local motion.

An object in a gravitational "pull" is actually:

- Sitting in a ripple slope
- Naturally aligning itself to lower ripple resistance
- This motion appears as falling

Why All Things Fall the Same

The beauty of UED becomes visible in a classic paradox:

- Why do heavy and light things fall at the same rate?

Because in UED:

- The gradient is in SVE — not in the object

- The object is merely resolving its position according to the ambient ripple imbalance
- Its internal mass doesn't change the slope — it rides the same path

Thus, gravity is:

- Mass-insensitive on the object end
- Mass-generated on the source end

We now see why:

- A feather and a hammer fall identically on the Moon
- There is no force "targeting" mass — only ripples seeking peace

Why Gravity Feels Attractive

Gravity feels like attraction. But it is no more "pulling" you to Earth than a slope pulls water. What really happens:

- The Earth is a massive ripple lock in SVE
- Around it lies a ripple tension field, with gradual alignment zones
- Any ripple-configured object (you, me, a stone) in that zone will naturally reconfigure toward equilibrium

This tendency manifests as motion — which we call "falling." What's missing in the traditional picture:

- No carrier particle (graviton has never been found)
- No quantum field working cleanly
- No continuous mechanism explaining *how* the force transmits

In UED:

There's no *transmission*. There's only resolution.

Why You Stick to the Ground

You're not "held down." You're sitting in a ripple minimum.

The Earth's ripple configuration creates:

- A gradient inward, toward its ripple center
- You, as a ripple configuration, are more stable closer to that minimum

That's why:

- You don't fly off the Earth
- Satellites orbit when ripple tensions are tangentially balanced
- Astronauts feel weightless when no net gradient acts upon them

Weight is not a force. Weight is your ripple misalignment cost.

Orbits Are Not Motion — They Are Alignment Locks

Take planetary motion:

- Newton called it gravity vs inertia
- Einstein called it curved geodesics
- NASA calls it escape velocity and centripetal force

UED says:

Planets are not moving. They are ripple-locked alignments within the gradient field of stellar SVE structuring.

Each celestial object:

- Occupies a ripple zone of stable tension
- Appears to move due to observer misinterpretation
- Is part of a dynamic equilibrium fabric

Hence:

- There is no "central force"
- The Solar System is not a mechanical machine
- It's a ripple-resolved sculpture in stillness

Why Gravity Is Always Attractive

Gravity never repels. Why?

Because ripple gradient always resolves inward toward equilibrium.

Other forces (like electromagnetic) can repel or attract because:

- Their ripple zones are localized and polar
- They can construct and deconstruct tension

But gravity is a field-wide equilibrium slope. It is structural — not interactive.

This explains why:

- Gravity can't be shielded
- Gravity works even across vacuum
- Gravity grows with mass, but not with complexity

It's not a "choice" — it's the default mode of resolution.

The Failure of Quantum Gravity

All modern attempts to quantize gravity:

- Graviton theories
- Loop quantum gravity
- String theory's curled dimensions

...fail because they assume force as a particle or exchange.

But UED states:

Gravity is not an interaction. It is a ripple asymmetry seeking balance. It cannot be *quantized* because it is not *emitted.*

It emerges only in massive context zones, where:

- Ripple configurations distort ambient SVE
- That distortion causes cascading realignments
- The objects inside the zone follow the path of least ripple resistance

Quantum mechanics is about interaction. Gravity is about contextual restoration.

They're not the same.

Gravity and Black Holes — A Coming Twist

In the next chapter, we'll see how UED kills the black hole.

Because:

- If gravity is just ripple gradient
- And ripple gradient flattens at extremes

Then the idea of an infinite well (singularity) breaks apart. There is no bottomless pull, only gradient exhaustion.

But for now, remember:

Gravity is not pulling you down. It's guiding you home — to the point of least tension.

Closing Reflection

There are no apples falling. No planets circling. No invisible threads pulling stars toward stars.

There is only one field — SVE. And it longs to return to silence. Gravity is just the field's whisper: "Come closer. Let's realign."

24: Black Holes Don't Exist

Something Else Does.

The Myth of the Monster and the Silence of the Structured Field

"You never saw a black hole. You saw structured silence — misunderstood as violence."

The Shadow that Science Worshipped

It was one of the most dramatic stories ever told in physics:

- A supermassive monster that devours everything.
- A point of infinite density where all known laws break down.
- A cosmic prison from which not even light can escape.

But in truth, this story was born not from observation, but from equations — specifically, mathematical extrapolations of general relativity.

We saw a shadow — and built a myth. UED invites us to see the structure, not the superstition.

The Origin of the Error: A Misread of Ripple Collapse

In the Universal Energy Dynamics (UED) framework:

- Mass is ripple configuration.
- Gravity is ripple gradient.
- Collapse is ripple disalignment and inward restructuring.
- Space and time are not fabric — they are ripple illusions.

So when a massive star "collapses" in traditional astrophysics, UED interprets it differently:

There is no singularity. There is no infinite gravity. What occurs is a ripple lock-in — a zone where the ripple field becomes so dense and still, it no longer interacts with external ripple configurations.

It's not a hole. It's a silence.

The Event Horizon: A Line We Drew on a Map of Confusion

Physics says the event horizon is the boundary beyond which nothing, not even light, can escape.

But light itself is a ripple resolution. And if ripple gradients are too steep, light fails to resolve visibly, not because it's trapped — but because it is not formed at all.

The event horizon is a perceptual threshold, not a physical wall.

We don't observe something "falling into" a hole. We observe the cessation of ripple reflection — the absence of light, not the presence of doom.

What the Telescope Actually Saw

The famous "black hole photo" — that fuzzy donut from the Event Horizon Telescope — wasn't an image of a black hole.

It was:

- A map of energy loss patterns
- A reading of interference zones
- A ripple shadow, not a monster

What we saw was not an object — but where structured ripple interaction failed to return visible feedback.

In UED terms:

- The center is a ripple equilibrium sink, silent to outer layers
- The glowing ring is where external ripple attempts still resolve

- The shape emerges not from pulling light in, but from alignment exhaustion

No Escape Velocity. No Infinity. Just Ripple Cancellation

The entire notion of a black hole requires:

- Infinite density
- Infinite gravity
- Finite escape velocity being exceeded

UED dismantles each:

- Density is a ripple construct. No structure can become "infinite."
- Gravity is not a pulling force, but a contextual gradient — and gradients flatten, they don't explode.
- Escape velocity is a Newtonian relic — ripple zones don't "escape"; they align or disalign.

A "black hole" is therefore:

Not a thing you fall into. It's a configuration where further ripple participation is no longer supported.

What Actually Exists: Ripple Condensates

Let's call it what it is: Ripple Condensate Zones (RCZs)

These are:

- Highly stabilized SVE structures
- With zero or near-zero ripple fluctuation
- Causing total non-reflection of observable energy
- Surrounded by zones of ripple exhaustion

They are:

- Dark (because no light resolves)
- Cold or silent (because no energetic interaction completes)
- Heavy in ripple tension, not in mass alone

They are not dead zones — they are perfectly structured, perfectly silent zones.

The True Cosmic Architecture

The universe isn't made of stars with occasional black holes.

The universe is a network of ripple configurations:

- Stars: intense ripple radiators
- Planets: ripple-balanced aggregates
- "Black holes": ripple silencers — not destroyers, but balancers

These ripple condensates:

- Anchor galactic alignment
- Regulate ripple density across cosmic zones
- Act as valves for SVE redistribution

Not monsters, but maintenance nodes of cosmic balance.

What Dies, and What Emerges

In standard theory:

- Stars die and become black holes.

In UED:

- Stars resolve their ripple structures
- When no further external or internal ripple participation is sustainable
- They collapse not into nothingness, but into a ripple-stilled state

This state is not "eternal" — it is maintained as long as surrounding ripple tension remains consistent.

RCZs can:

- Dissolve
- Expand
- Reconfigure — based on ambient ripple context

The Absence of Time and the Illusion of Singularity

Time ceases to function in the black hole, says relativity.

UED says:

Time never "existed" there to begin with. Ripple gradient flatlines — and without gradient, no change is measured. Hence, time is not halted, just never born in that context.

The idea of a "singularity" — a point of infinite curvature —is merely a placeholder for our confusion.

Ripple dynamics never go infinite. They self-resolve or disalign — they don't self-destruct.

Closing the Hole, Opening the Field

So what exists in place of a black hole?

- No hole
- No singularity
- No point of no return

But instead:

- A ripple condensate, silent and structured
- A tension sink, not a gravitational trap
- A non-interfering node of SVE balance

In short:

The quietest part of the universe — not the loudest. Not a roaring maw, but a whispering core of cosmic stillness.

Closing Thought

The hole was never real. What you saw was the mirage of misperception, projected on the silence of perfect structure.

In the end, the deepest truth wasn't monstrous. It was peaceful — waiting to be understood without fear.

25: Light, Waves, and Energy

What They Really Are

From Ripple Disturbance to Consequential Resolution

"You never saw light. You witnessed resolution. You never stored energy. You balanced misalignment. You never felt a wave. You felt the universe trying to settle itself."

Unwrapping the Illusions

Modern science taught us that:

- Light is both a wave and a particle.
- Waves can travel through nothing.
- Energy is a thing that can be stored, transferred, and converted.

Yet behind these terms lies a tangle of unresolved contradictions.

UED doesn't offer another metaphor. It offers clarity: Everything—light, wave, energy—is a ripple consequence, not a cosmic object.

Light: Not a Particle, Not a Wave — A Ripple Resolution

Let's start with light.

In physics:

- Light travels at 299,792,458 m/s in vacuum.
- It's modeled as a photon (massless particle) and a wave.

- It can reflect, refract, diffract, and interfere.

But all these properties only appear when:

- A ripple field interacts with structure (material or energetic)
- Resolution occurs — meaning: the SVE pattern aligns and delivers a perceivable outcome

In UED:

- Light does not travel like a projectile.
- It is a resolved consequence between two aligned ripple zones.
- No photon *flies* from a star to your eye — instead, structured ripple pathways resolve visual experience when alignment is achieved.

Light is not motion. It is momentary clarity in the ripple continuum.

Waves: Not Things, But Disturbances Seeking Equilibrium

A wave in classical science is often imagined as a thing moving through a medium. But this imagination fails in space. What moves in a vacuum?

In UED:

- There is no vacuum. There is only SVE—structured or unstructured.
- A wave is a temporary ripple imbalance.
- It represents the system's attempt to return to equilibrium.
- The frequency is how often the ripple tries.
- The amplitude is the intensity of imbalance.

Waves don't move. Ripple disturbance moves forward through resolution. What appears as propagation is sequential resolution — not transport.

The wave you see is just the universe whispering, "I'm not done yet."

Energy: The Most Misunderstood Concept in Science

What is energy?

We're told:

- It's the capacity to do work.
- It exists in many forms: kinetic, potential, thermal, nuclear, etc.
- It can neither be created nor destroyed.

Yet:

- We've never seen it.
- We can't isolate it.
- We only infer it from consequences.

UED redefines energy:

Energy is ripple configuration in active resolution.

It is:

- Not a stored commodity, but a tension state in the ripple field.
- Not an object, but a degree of disalignment in the SVE structure.
- Not transferable, but redistributable via ripple propagation.

For example:

- A stretched spring has potential energy because its ripple field is tense.

- When it's released, the ripple field realigns, and we call that kinetic energy.
- No "thing" moved — only pattern resolution occurred.

Photons and Quantum Misinterpretations

In quantum physics:

- A photon is a discrete packet of light.
- It exhibits wave-particle duality.
- It "collapses" into particle state upon observation.

But UED says:

- There is no particle.
- There is no wave.
- There is only structured ripple alignment.
- "Collapse" is simply the completion of ripple negotiation.

What scientists call photon detection is merely:

- A ripple match between the emitter and detector.
- Once alignment occurs, the field stabilizes, and resolution happens.
- There is no transfer, only completion.

Energy Is Always Local

Mainstream physics says energy can travel across:

- Distances
- Vacuum
- Infinite scales

UED corrects this:

- Energy is local ripple behavior.
- What seems like "far transmission" is really re-alignment through intermediate ripple nodes.
- This is why interference, attenuation, and collapse occur — because the ripple can't maintain alignment over distance unless continuously resolved.

Think of wireless charging or radio waves:

- Nothing travels.
- Ripples interact across potential fields until alignment is achieved.
- The "field" is not an entity — it is a ripple tension network.

Why We Feel Heat, Light, Sound — Yet Never Energy Itself

We say we feel:

- Heat from the sun
- Sound from a speaker
- Light from a lamp

But we never feel "energy." We only perceive resolved ripple interaction.

Every sensory experience is:

- A resolved ripple path
- From external source to internal structured pattern
- Processed through ripple-activated perception frameworks (i.e., the Mind)

"Energy" is the label for what happens between cause and consequence, not the thing in itself.

Realizing What Energy, Wave, and Light Actually Are

Concept	Mainstream View	UED View
Light	Particle-wave duality	Ripple resolution through structured alignment
Wave	Object moving through space	Pattern of ripple imbalance seeking equilibrium
Energy	Transferable, storable entity	Local ripple tension state —redistributed, not transferred
Photon	Massless particle/quanta	Momentary ripple match —not a thing, but an event

The Consequence Framework: A Final Word

Everything we define:

- Mass
- Force
- Light
- Time
- Energy

...is not fundamental. It is consequential — a byproduct of ripple dynamics within SVE.

UED doesn't reduce the world. It reveals it.

No wave ever traveled. No energy was ever stored. No light ever moved. The universe simply aligned — and we called that reality.

Closing Reflection

"In the beginning, there was stillness. Then came imbalance. And what you now call energy, light, and wave —were simply the ripple's way of coming home."

26: Why the Universe Was Never Expanding

The Stillness That Appears to Grow

"You're not watching space expand. You're witnessing resolution ripple outward— not in distance, but in depth of alignment."

The Great Assumption: Expansion from Observation

Mainstream cosmology claims:

- The universe began with a Big Bang—a singularity that exploded into space.
- Since then, space itself is expanding.
- Galaxies are moving away from us, and redshift proves it.
- The cosmic microwave background (CMB) is an echo of this explosive beginning.

This idea rests on one assumption:
That distance change = motion, and motion = expansion.
But in UED:

- There is no motion as imagined.
- There is no space fabric stretching.
- There is no bang to begin with.

There is only:

Ripple alignment unfolding across still SVE — giving the illusion of expansion when observed locally.

The Universe is Not a Balloon

The popular analogy:

- Imagine the universe like a balloon inflating.
- Galaxies are dots on the surface—moving apart not because they're traveling, but because space itself is growing.

This balloon idea fails under UED:

- SVE is already infinite — there is no "edge" to stretch.
- There is no medium expanding—only local ripple interactions resolving over new configurations.

Expansion is not a physical change in volume. It is a consequence of ripple access broadening, not space increasing.

Redshift: Not Evidence of Expansion — Evidence of Resolution Delay

In classical cosmology:

- Light from distant galaxies is redshifted—its wavelength stretched.
- This is taken as proof that those galaxies are receding from us.
- Greater redshift = greater distance = faster recession.

But UED reframes this:

- Redshift is not due to motion.
- It is due to ripple misalignment between source and observer.
- The further the source, the more ripple noise or distortion accumulates.
- The delayed resolution appears as stretched wavelength (redshift).

You're not seeing objects move away. You're watching resolution lose clarity — and we called it speed.

Cosmic Microwave Background: The First Resolved Layer, Not a Big Bang Echo

The CMB is often called:

- The "afterglow" of the Big Bang.
- Radiation left over from the early universe, when light first escaped opaque plasma.

But UED says:

- The CMB is not a leftover.
- It is a baseline ripple pattern that resolved across the first stable alignment zone.
- It exists not because the universe cooled — but because ripple coherence achieved its first long-scale resolution.
- It is not past light—it is present structure.

There Was No Bang — There Was a Ripple

What caused the illusion of beginning?

Mainstream says:

- Everything came from nothing in an instant.
- That nothing exploded due to infinite density.
- Time and space were born in that moment.

UED corrects:

- There was no explosion.
- There was a ripple — the first tension in otherwise still SVE.
- Time is what emerged from that ripple.

- Structure unfolded—not outward, but inward, through layers of alignment.

Creation was not an event. It was a sequence of resonance — a song, not a bang.

Expansion vs Expression

Here is the core difference:

Concept	Mainstream View	UED View
Expansion	Increase in spatial distance	Increase in ripple resolution and access
Motion	Objects flying outward	Ripple interaction resolving locally
Redshift	Receding galaxies	Consequential signal misalignment
CMB	Past radiation from Big Bang	Current baseline ripple layer
Origin	Big Bang singularity	First ripple from stillness (SVE)

The illusion of motion is not real expansion. The increasing visibility of structured ripple layers creates a dimensional mirage.

You feel like you're seeing the universe "grow" because:

- Your alignment with its structured ripple field is deepening.
- Perception bandwidth is expanding — not the universe.

What Appears Far Is Simply Less Aligned

Distant galaxies are not far in space. They are far in resolution. This explains:

- Why we never "reach" the edge of the universe.
- Why the farther we look, the older and redder things seem.
- Why the expansion rate appears to accelerate (we're misreading ripple time effects).

There is no "outward rush." Only a ripple field resolving in shells of coherence. The universe didn't expand. You arrived deeper into its awareness.

The Still Universe — Not Static, Not Expanding, Just Aligned

UED offers the middle path between:

- The static universe (Einstein's initial idea),
- And the expanding universe (Hubble's misread data).

It proposes:

- A still universe with evolving alignment.
- A cosmos that doesn't move, but becomes increasingly coherent.
- No need for dark energy, inflation, or fabricated constants.

Every ripple layer you discover is:

- Not a frontier, but a mirror reflecting how deeply your own structure aligns with universal structure.

You are not traveling through space. You are unfolding through alignment zones.

Final Reflection:

What we thought was expanding was just us— resolving into the ripple field that was always there.

The beginning was not the Big Bang. The beginning was stillness...finally disturbed ...by a ripple that remembered its source.

27: Dark Matter & Dark Energy

Shadows of Misalignment

The Ghosts That Never Were

"You can't find what was never structured. But you can resolve what was always misunderstood."

The Crisis That Gave Birth to Ghosts

Modern science didn't invent dark matter and dark energy out of curiosity. It invented them out of confusion.

- When galaxies rotated faster than they should based on visible mass, physicists postulated extra, invisible mass: *Dark Matter.*
- When the universe appeared to be expanding faster over time, they invented a repulsive force to account for it: *Dark Energy.*

These were not discoveries. They were corrections to equations that were breaking.

Dark matter and energy are not physical entities. They are placeholders for ignorance—desperate attempts to make flawed models match real data.

What Are They Supposed to Be?

Dark Matter:

- Makes up 27% of the universe (as per standard model).
- Doesn't emit, reflect, or absorb light.
- Inferred only through gravitational effects—like galaxy

rotation curves and gravitational lensing.

Dark Energy:

- Accounts for 68% of the universe.
- Drives the "accelerating expansion" of space.
- Inferred from Type Ia supernova data and redshift interpretation.

Together, these "dark" components supposedly form 95% of all that exists.

But here's the twist: None of them have ever been directly detected. Not a single dark particle. Not a single dark field.

They exist only in calculations—and in the minds of those who fear letting go of Newtonian-Einsteinian assumptions.

What Really Went Wrong? (UED View)

The failure is not in observation. It is in interpretation.

Flawed Assumptions:

1. Space is expanding.
2. Gravity is a force acting between masses.
3. Light is stretched by velocity or space itself.
4. Time is linear and uniform.
5. Every interaction is local and mass-based.

These are the foundations upon which dark matter and energy were proposed. UED challenges each one.

UED Reframes the Entire Scene

In Universal Energy Dynamics (UED):

- There is no "space" that expands.
- There is no motion — only ripple alignment and delay.
- Mass is not substance — it is ripple configuration.

- Force is not an entity — it is a consequence of misalignment.
- Gravity is a gradient of ripple anchoring — not attraction.

The entire cosmos is still. What we perceive as movement or expansion is ripple alignment appearing over time through different resolution layers.

The Truth Behind Dark Matter

Problem: Galaxies rotate faster than their visible mass predicts. They should fly apart — but they don't.

Mainstream Answer: There must be hidden mass holding them together.

UED Answer:

- No hidden mass exists.
- Galaxies are not held together by gravity alone.
- They are locked into place by their ripple interface with the Structured Vacuum Energy (SVE).
- What appears as "mass effect" is actually the resistance of ripple dissipation across structured SVE.

Think of it like a spoon moving through thick honey. It moves differently than in air — because of the medium's structure.

Galactic behavior is not wrong. Our model of mass is. What they called "dark matter" was actually the ripple tension of invisible structure.

The Illusion Behind Dark Energy

Problem: Distant galaxies appear to recede faster over time. This implies acceleration in the universe's expansion.

Mainstream Answer: There must be a mysterious repulsive energy — "dark energy" — driving everything apart.

UED Answer:

- There is no outward motion.
- Redshift is not due to space expanding.
- It is due to resolution decay — the further we "look," the less aligned the ripple structure is between observer and source.
- The so-called acceleration is a perceptual artifact caused by ripples weakening across distance.

In other words:

You are not watching galaxies rush away. You are watching clarity fade across the ripple field.

Why They'll Never Be Found

No particle detector, no telescope, no mathematical tweak will ever "discover" dark matter or dark energy. Because they are not things. They are misread phenomena—symptoms of a model that doesn't account for SVE or ripple alignment.

Their continued invisibility is not a mystery. It's a verdict.

"They don't show up in experiments because they don't exist in structure."

The Real Explanation Table

Phenomenon	Mainstream Fix	UED Reality
Galaxy rotation speed	Dark Matter	Ripple anchoring in SVE density zones
Accelerating expansion	Dark Energy	Delay in ripple alignment resolution
Gravitational lensing	Mass bending space	Ripple redirection across misaligned ripple zones
Uniform CMB despite scale	Inflation & DE	Baseline ripple coherence across structured shell
Failed particle detection	Undetectable fields	There are none — only consequences of misreading

The Final Reframing

There are no dark entities pulling or pushing the cosmos.

There is only:

- A ripple field resolving through still SVE.
- Perception delayed by misalignment.
- A scientific tradition built on consequence before cause.

UED corrects this. It doesn't add more variables. It removes the illusion.

Final Reflection

They searched the void, for what was never missing. They invented ghosts, because they didn't see the pattern.

Dark matter is what happens when you forget the structure. Dark energy is what happens when you misread the stillness.

The universe was never dark. It was always too clear for our old eyes.

Interlude: When Science Looked Again

and Saw What Was Always There

"The universe didn't change. Only the way we understood it did."

This part wasn't a revision. It was a liberation. We didn't just rewrite equations. We unwrote assumptions.

A Summary of Shifts:

1. Time is not ticking — it is consequence, not container. It is the rhythm of ripple unfolding, not an ever-moving line.
2. Mass and Charge are not fundamental things — they are configurations. Arrangements of ripple density and gradient—temporary roles in a cosmic play.
3. Electromagnetic force is not about invisible lines or abstract fields. It is ripple agreement between zones seeking coherence.
4. Gravity is not attraction. It is the gradient formed when structure resolves into tension.
5. Black holes? Not monstrous pits of destruction. Just the regions of unresolved ripple convergence, where illusion deepens to boundary.
6. And no — the universe was never expanding. Our perception was.

The fabric didn't stretch — we mistook misalignment for motion.

What We Truly Did

We peeled the mathematical masks off the cosmos. We replaced fantasy with structure. We resolved illusion with intelligence.

We stopped chasing the symptoms — and started listening to the source.

A New Science Begins Here

- No "dark" matter.
- No singularities.
- No expanding voids.
- No gravitational fabric, bent and bruised.

Only a still, intelligent field. A universe made of ripple consequence — measurable, meaningful, and no longer mysterious.

We didn't need more answers. We needed to ask again, cleanly. And when we did, we didn't just get new science.

We got the science of else. The science that was waiting — quietly — beneath every wrong question ever asked.

Now, the Real Game Begins

What happens when this new science meets the idea of God? What if this ripple field wasn't just physics...but the mechanics of meaning itself?

What if every law we followed was never written in equations...but embedded in alignment?

In the next part, we explore the Divine Mechanics and the Law of Else.

Because now that we've resolved science,we're finally ready to resolve Self and Source.

Part VI: Spirituality Without Symbols

28. God Is SVE

Stillness, Not a Skyfather

God is not someone. It is the structured stillness from which all consequences arise.

For millennia, humanity has imagined God in form — often as an old, wise figure in the clouds, watching, rewarding, punishing, intervening. This image was emotionally useful, culturally powerful, and politically convenient. But now, with clearer understanding through Universal Energy Dynamics (UED), we must evolve this image — not to discard divinity, but to reframe it as pure intelligence in stillness.

- God is not above.
- God is not waiting.
- God is not judging.

God is the structured stillness — the foundational state of the universe before consequence, the SVE (Structured Vacuum Energy) field.

Why the Old Image Was Never God

The Anthropomorphic Trap

- Religions often describe God in human terms — with emotions, anger, love, preferences.
- This made God relatable but also limited — turning the infinite into a character.

- This led to contradictions: *Why does a loving God allow suffering? Why would a perfect being want worship?*

God as Mythical Manager

- The "Skyfather" image makes God an external agent — someone who created, then governs from afar.
- In this framework, humans are passive recipients: "good or bad" based on divine moods.
- But this removes responsibility from the individual — outsourcing both consequence and wisdom.

The Need for a New Understanding

- As science advanced, this old model began to collapse. But science didn't replace God — it just left a vacuum.
- UED doesn't discard God. It resolves God.

Stillness Is the Source of All

What Is SVE?

Structured Vacuum Energy is not emptiness. It is perfect, motionless, silent potential. It is still — yet all ripples emerge from it. It is neutral — yet every configuration finds its equilibrium within it. It is not a being, but the foundation of being.

Stillness, Not Absence

Stillness is often misunderstood as "nothing." But in UED, stillness is everything before it becomes anything.

- It holds no bias but allows all forms.
- It doesn't speak, but it echoes through alignment.
- It doesn't create with intention — but creation happens through misalignment and resolution within it.

The True Divine

What we once called "God" is this ever-present, unchanging stillness.

- Not controlling, but containing.
- Not rewarding, but resolving.
- Not fearing, but flowing through all things.

This is not a lesser version of God. It is a more complete one — without story, without symbol, yet utterly sacred.

Positive Reframing — What It Means to Believe in God Now

No More Fear

You are not being watched or judged. Stillness does not punish. Every consequence arises from ripple misalignment — not divine wrath. This shifts us from a fear-based morality to awareness-based action.

Real Relationship: Not Prayer, but Presence

You don't have to "reach" God. You are already made of ripple aligned with SVE. To know God is not to shout at the sky — but to resolve into stillness.

You don't find God. You stop creating noise, and God — stillness — is what remains.

From Devotion to Alignment

Ritual is optional. Worship is not mandatory. But alignment is everything.

- When your thoughts align with clarity,
- When your actions align with consequence,
- When your mind quiets into the field —

You are in communion with the divine.

Scientific Spirituality — God Measurable, Not Mystical

SVE Is Not Speculative

In UED, SVE is not a metaphor. It is the structured substrate. It has implications in physics, consciousness, death, emotion, and perception. It interacts through ripples and defines what exists, what dissolves, and what aligns.

Divine Intelligence, Not Superstition

This is the science of God — not myth-based, not faith-based, but structure-based.

- The Law of Consequences (karma) arises from SVE structure.
- The soul is a structured ripple zone.
- The mind is ripple awareness in action.
- Death is ripple collapse back into SVE.

All of it fits. Nothing contradicts.

God Within, Not Above

"You are not a creature looking up at God. You are the ripple watching itself in the stillness it came from."

- You don't need intermediaries.
- You don't need dogma.
- You are never separate from divinity.

Once you stop believing in God as character, you start experiencing God as condition.

And that condition is:

- Still.
- Pure.
- Universal.

- Intelligent without intention.
- Divine without division.

Closing Reflection:

God is not the cause. God is the context in which causes resolve. Not "He." Not "It." Not "They." But That — which remains when all else dissolves.

There is no need to imagine God anymore. When you finally align — You won't have to.

29. Karma: Not Justice, But Consequence

The ripple never rewards or punishes. It only continues.

The word *karma* carries more baggage than clarity. To most, karma means justice — a cosmic scoreboard where good deeds earn rewards and bad deeds bring suffering. This belief comforts some and frightens others. It has been used to inspire morality... and to justify misfortune.

But what if this whole framing is flawed?

In UED, karma isn't cosmic morality. It is cosmic mechanics.

Karma is not judgment. Karma is not fairness. Karma is simply consequence — ripple continuation in the structured field.

Karma as Mechanics, Not Morality

What the Ripple Really Does

Every action — thought, word, movement — produces a ripple in the Structured Vacuum Energy (SVE). These ripples interact with existing structure. They create alignment, friction, resonance, or distortion.

The universe doesn't judge those ripples. It doesn't decide whether your action was good or evil. It only processes the outcome.

And once the ripple is produced, it must resolve — in you, around you, or through time.

Karma is not a record. Karma is a ripple timeline unfolding in real-time.

There Is No Karmic Accountant

- No divine being is tallying points.
- No "bank" of good or bad energy.
- No rewards or punishments.

There is only the field — structured and precise — and your interactions with it.

When aligned, the ripple fades peacefully. When misaligned, it rebounds, distorts, or gets entangled in further consequences.

The Illusion of Justice

Why Good People Suffer

In the justice model, we ask: *"Why do bad things happen to good people?"* But the ripple doesn't care about labels. "Good" is a human emotion. The ripple only cares about configuration.

A person could be kind-hearted yet unaware of how their actions misalign with surrounding structure. The resulting ripples cause effects — not as punishment, but as consequence.

Similarly, someone manipulative may appear to succeed, but their ripple network becomes increasingly unstable. Collapse is not punishment. It's inevitable misalignment.

Karma Is Not Immediate

Another common misunderstanding: karma must be quick. But ripple consequences are not always local or instant.

They may:

- Reflect back later (in the same life),
- Influence others around you,
- Resonate through systems and institutions,
- Continue across generations via societal structure.

This isn't mystical. It's systemic resolution. And delayed consequence is still consequence.

Self, Karma, and Responsibility

You Create the Ripple

You don't carry karma. You generate it, moment to moment.

- Every thought you entertain.
- Every intention you plant.
- Every action you commit.

They all emit ripples. And the universe, like a still lake, responds accordingly.

This isn't about guilt. It's about awareness. You are not under karma's rule. You are its architect.

Karma is not fate. It is your interaction history with the universe.

Stop Outsourcing Karma

Stop saying:

- "This must be my bad karma."
- "She's suffering because of her karma."
- "The universe will take care of them."

These are misunderstandings — often used to avoid responsibility or feel righteous.

Karma is not revenge outsourced to nature. It is feedback, not justice.

UED's Law of Consequences (LOC)

UED defines karma through a scientific lens:

- Ripple is the medium of interaction.
- Alignment defines sustainability.
- Every misalignment produces an effect.
- That effect continues until resolution.

This is the Law of Consequences (LOC) — the universal mechanism by which energy, intention, and structure respond to each other.

No belief is required. No ritual needed. You are already participating.

Just as gravity doesn't need your permission to act, karma (or LOC) doesn't need your belief to continue.

Karma and Liberation

Can You Escape Karma?

No — and yes.

You can't escape consequence, because to live is to ripple. But you can align. You can learn to act without distortion. You can dissolve karma through awareness + right configuration.

This is not about asceticism or passivity. It's about becoming consequence-aware — seeing how even your smallest ripple shapes reality.

Action Without Attachment — Now Reframed

The Gita speaks of *karma-yoga* — action without attachment. UED reframes it as: Ripple with minimal distortion.

You still act. But with:

- Clarity, not compulsion.
- Awareness, not anxiety.
- Precision, not projection.

Final Reflection: No One Is Watching. Everything Is Responding.

There is no cosmic judge. There is no fate-maker.

There is only structure. And your ripple inside it.

Karma is not waiting for you after life. Karma is right now — in every interaction. Not as reward. Not as punishment. But as consequence that teaches.

You are not bound. You are responsible. And that makes you powerful.

30. Prayer Is Not Asking — It Is Aligning

Real prayer doesn't beg. It synchronizes with the field that already knows.

For centuries, prayer has been portrayed as a plea — a spoken or silent request to a higher power, an emotional outreach to the heavens for help, mercy, or intervention.

We bow. We fold our hands. We whisper or cry aloud. We *ask* — and we *hope* someone is listening.

But who exactly are we asking? And where is this "God" who responds?

The Traditional Model of Prayer

Petition-Based Framework

Most major religions treat prayer as a transaction:

- Input: Supplication, praise, confession, request.
- Expected Output: Blessing, guidance, healing, reward.

This framework assumes:

1. God is a being with emotions and moods.
2. That being has free will and may choose to grant or deny.
3. Humans are powerless without divine approval.

In this model, prayer becomes pleading — often mixed with guilt, fear, and doubt.

But this view collapses under deeper observation — and under UED.

The Reality of Prayer in UED

In the Universal Energy Dynamics (UED) framework, there is no Skyfather. There is no being "out there" who needs to be convinced.

There is only Structured Vacuum Energy (SVE) — the intelligent stillness from which all ripple arises.

You are not separate from this field. You are a ripple localized within it.

So when you pray, you are not sending a signal outward. You are reconfiguring inward — into alignment with what always was.

Prayer is not a call. It is a correction. A self-tuning. A field recalibration.

What Happens When You Truly Pray

When you stop begging and start aligning, prayer becomes an act of conscious integration.

Here's what shifts:

1. From Asking to Listening

You no longer speak "to" God. You listen into the silence — because the structure already knows.

You quiet the ripple distortions (fear, doubt, desperation) and allow the original stillness to guide reconfiguration.

This isn't passive. It's receptive awareness — a powerful act.

2. From Hope to Harmony

Old prayer: *"Please fix this."* Aligned prayer: *"Let me see clearly."*

You're no longer trying to control outcomes. You're tuning yourself to act with minimal consequence distortion.

It's not about surrendering power. It's about merging with the source of all power.

3. From Reward-Seeking to Ripple Coherence

You don't pray for miracles. You pray for resolution — internal and external ripple harmonization.

That's the miracle.

Scientific Mechanism Behind Prayer

If you see the body as structured SVE, and the mind as ripple interpretation...then prayer becomes a ripple-resolving protocol.

You align mind → emotion → breath → body → intent into one coherent configuration.

This coherence:

- Stabilizes internal fields (emotionally and physically),
- Reduces mental noise,
- Allows ripple signals from the SVE to re-emerge without distortion.

That's why prayer is often followed by:

- Sudden insight,
- Emotional relief,
- Unexpected clarity or calm.

The field never changed. You did.

What Real Prayer Looks Like

Let's reframe prayer through practice, not performance:

1. Stillness

Sit. Stand. Lie down. Doesn't matter. Just stop moving internally.

Stillness is the invitation for original structure to reappear.

2. Presence

Drop the story. Drop the request. Drop even the idea of "God."

Become fully here.

This allows the ripple cloud of identity and emotion to settle.

3. Intention (Pure, Not Desperate)

Ask without need. Desire clarity, not control. Want truth, not escape.

The purest prayer is:

"Let what must resolve, resolve — and let me be aligned enough to see it."

4. Gratitude (as Alignment, Not Obligation)

Not for what you "received." But for the fact that you could feel the field again.

Gratitude is recognition of stillness, not reward.

Prayer Beyond Religion

This version of prayer belongs to no religion. It requires no god-name, temple, scripture, or posture.

- A scientist can pray through inquiry.
- An artist through immersion.
- A teacher through honest presence.
- A parent through aligned love.

Prayer is whenever ripple becomes aware of its origin — and synchronizes with it.

You don't need belief to pray. You need willingness to align.

Final Reflection: God Already Knows

The field doesn't need your explanation. It doesn't need your performance.

It just waits for your alignment — patiently, precisely.

Real prayer isn't heard. It is felt — by the universe as you fall into coherence with it. And in that moment, nothing needs to be said. Because everything is already resolved.

31. Meditation: Collapse into Causeless Being

Silence is not emptiness — it is where all ripple collapses, and origin is remembered.

Meditation, like prayer, has long been branded, packaged, and misused.

It is often taught as a tool for:

- Stress relief
- Focus enhancement
- Emotional regulation
- "Spiritual growth" or even "manifestation"

But these are by-products, not its essence.

True meditation is not an act. It is a collapse — a resolution of all consequences into the source that precedes them.

Meditation is not doing. It is undoing.

Ripple and the Illusion of Identity

Every thought you think, every feeling you feel, every desire you chase — is a ripple in Structured Vacuum Energy (SVE).

Ripples are not bad. They are how the still field *expresses* possibility and consequence.

But with time, we forget the stillness behind the ripple. We start believing:

- "I am the mind."
- "I am the voice inside."

- "I am the story unfolding."

Thus begins the illusion of being someone — a separate self, living a linear life.

Meditation is the undoing of that illusion.

What Truly Happens in Meditation

When you meditate in the UED sense, you are not thinking about truth.

You are realigning with it.

Let's break this down:

1. Ripple Slowing

The chaotic ripples of thoughts, identity, and emotions begin to lose momentum.

This is not suppression — it's resolution.

Just as a disturbed water surface returns to stillness without force, your mind returns to its original field if not fed by reaction.

2. Boundary Dissolution

As ripples dissolve, the illusion of "I" — the rigid edge of ego — also softens.

There is no longer "you meditating" and "something else being observed."

The field becomes non-dual again — observer, observed, and observing merge.

3. Collapse into Source

The final phase is not transcendence upward — but collapse inward.

Not expansion, but *reduction of consequence*. Not escape, but entry into the uncreated.

This is causeless being.

You are not "feeling peace." You are becoming the space in which even peace disappears.

Why Silence Is Not Emptiness

Many fear silence. They say it is boring, dark, lifeless.
But this is only because they are addicted to ripple noise.
Silence is not the absence of sound. It is the presence of origin.
In deep silence:

- Time does not tick.
- Thought does not narrate.
- Self does not demand.
- The universe does not explain.

And yet — everything is right.
This is the paradox of SVE:
That which appears as nothing... is the source of all somethings.

Meditation Is Alignment, Not Escape

Meditation is often sold as:

- A way to detach
- A method to "rise above" the world
- A tool to "manifest" desires

But this is spiritual consumerism — not realization.
True meditation doesn't add power. It removes distortion.
You don't get stronger. You become clearer.
You don't escape the world. You return to the structure behind it.
In that structure, desire dissolves. There is nothing to gain — because you are back at source.

Scientific Mechanism of Meditation (UED Perspective)

Here's how meditation works under Universal Energy Dynamics:

- Mind = ripple activity
- Emotions = ripple distortions

- Body = structured ripple consequence
- Suffering = misalignment between ripple layers
- Meditation = ripple collapse and re-synchronization

During deep meditation:

- Ripple activity reduces,
- Internal gradients equalize,
- Structured zones return to coherence,
- Observer merges with origin.

It's not altered consciousness. It's returned structure.

And this return heals — not because it adds anything, but because it removes interference.

How to Meditate (Without Rituals)

No chanting required. No postures. No belief system.

Just stillness and allowing.

Step 1: Sit in any position. Eyes open or closed.

Let your body be still — but don't resist its natural movements.

Step 2: Do not* focus. *Just observe.

No need to count breaths or repeat mantras. Just notice what arises and dissolves.

Step 3: Don't hold. Don't react.

Thoughts will come. Let them pass like waves. Emotions may surge. Let them pass like clouds.

Step 4: Allow awareness to soften.

Let the boundaries between self and space blur. Let the inner chatter become background noise.

Step 5: Wait. Don't expect.

Eventually, ripples will reduce. Stillness will surface. And something deeper will begin to remember itself — You, before you became "you."

Final Reflection: The Original State

You weren't born from chaos. You arose from stillness.

Before thought, memory, desire — You were SVE in ripple form.

Meditation is how the ripple re-meets its source. Not to worship it. Not to decode it.

But to become it, again.

Meditation is not awareness of stillness. It is stillness becoming aware — of itself. And in that, Nothing remains unresolved. And nothing ever needed to.

32. Destiny, Luck, and Signs

No Magic Needed

What feels divine is often ripple coherence. The universe speaks, but not in miracles.

The idea that the universe operates on divine will, hidden signs, or magical fate is one of the most enduring beliefs in human civilization.

- A bird flies overhead when you're thinking about someone: *a sign.*
- You narrowly miss an accident: *luck.*
- A dream predicts an event: *destiny.*

But what if none of these are *supernatural*?

What if the so-called "mystical coincidences" are simply outcomes of a coherent ripple structure, operating according to natural laws beyond the mind's usual comprehension?

In the UED framework, the universe does not operate through miracles. It operates through alignment — of energy, fields, observers, and consequences.

Let's dive in.

What Is Destiny?

Destiny is often confused with:

- A fixed future

- A divine plan
- An unavoidable script

But under UED, destiny is not what's written — it's what aligns.

Let's define it precisely:

Destiny is the set of probable consequences that emerge from the structured alignment of ripple fields — across time, intention, and interaction.

You are not *fated* to become something. You are *structured* by your choices, your environment, and the ripple fields you engage with.

Destiny is dynamic.

It is not pre-written. It is pre-structured — based on what you've aligned with so far.

It appears fixed only when you:

- Continuously repeat misaligned patterns,
- Refuse to recalibrate,
- Or remain in resonance with distorted environments.

Thus:

- Good destiny = consistent ripple alignment with stable fields
- Bad destiny = sustained misalignment from natural coherence

It's not God playing chess. It's the ripple field resolving its own imbalances.

What Is Luck?

Luck is often seen as:

- Favor from the divine
- Random good fortune
- A force beyond control

But in ripple physics, "luck" is neither random nor divine.

It is simply the unexpected resolution of ripple friction — a sudden, favorable alignment that feels improbable to the untrained mind.

For example:

- You meet the right person at the right time → Luck? No. Ripple convergence.
- Your idea gets viral suddenly → Luck? No. Resonance + amplification in an unseen field.

So:

Luck = Invisible coherence resolving in your favor.

It's invisible only because:

- You don't see the field gradients,
- You don't track ripple history,
- You misinterpret consequence as coincidence.

In UED, luck is not a gift — it's a consequence of silent structural momentum.

You "make your own luck" not by wishing, but by:

- Living in alignment,
- Removing internal friction,
- Maintaining resonance with larger coherent systems (people, purpose, environment).

What About Signs?

From ancient omens to modern angel numbers, humans have long believed that the universe sends signs. And in a way, it does. But not in the magical sense.

Here's how to reframe it:

A "sign" is a ripple echo — a consequence from the field that becomes meaningful when your awareness aligns with it.

It's not that the universe is *talking to you*. It's that:

- You are aligned enough to notice subtle ripple intersections,
- Your inner state matches a larger field structure,
- And therefore what you see feels profound, not because it is magical, but because it is structurally valid.

Think of it this way:

- A compass always points north — not by divine instruction, but due to field coherence.
- Your perception aligns to "meaningful coincidence" when your internal compass is undistorted.

That's when signs appear.

The Universe Is Coherent, Not Magical

Mainstream spirituality often says:

- "Everything happens for a reason."
- "The universe has a plan."
- "Miracles are real."

UED corrects this with elegance:

Everything happens for a consequence, not a reason. The universe has structure, not a plan. And miracles are just misunderstood ripple alignments.

Let's break the myth of miracle:

Water turning to wine? In ripple physics, molecular restructuring can be induced by field manipulation — but that's science, not sorcery.

Knowing future events? Memory of ripple flow, not prophecy.

Healing through touch or intent? Coherence-induced ripple balancing, not magic.

Miracles are edge phenomena — field-level interactions that appear magical only because we haven't resolved the layers yet.

Aligning with "Fate" the UED Way

Instead of asking:

- "What is my destiny?"
- "When will luck favor me?"
- "What does this sign mean?"

Ask:

- "Am I aligned with stable ripple structures?"
- "Are my internal fields coherent with what I seek?"
- "Am I acting in resonance or resistance?"

This shifts you from passive hope to active alignment. Because:

- Destiny is not imposed. It is resolved.
- Luck is not luck. It is unnoticed alignment.
- Signs don't speak. You *remember* them when you're ready.

Practical Reframe — Living Without Superstition

Here's how this insight transforms life:

Old Paradigm	Ripple-Aligned Reframe
"It's all written in fate"	Your alignment writes your outcome
"Bad luck keeps following me"	Misalignment is attracting repeating patterns
"I saw a sign from the universe"	You perceived a coherent ripple intersection
"Some people are just lucky"	Some people live in ripple alignment
"Why me?"	Why not? You're part of the same ripple field

This isn't to deny wonder or awe — in fact, it enhances it.

Because now, the beauty of life is not in divine favoritism, but in universal fairness.

Every ripple matters. Every choice changes the field. And you are never too far gone to realign.

Final Reflection: No Magic. Just Marvel.

Once you stop expecting angels to land or stars to spell messages...

You begin to see that the real miracle was always:

- The coherence of the field
- The grace of realignment
- The gentle nudges of consequence

You stop begging for signs. And start becoming one — in someone else's ripple path. Don't follow stars. Be one.

Interlude: The Sacred Without the Story

There was a time when humans needed myths to explain thunder, stories to explain stars, and gods to explain fear. We built temples to externalize what was internal, drew symbols to make stillness visible, and prayed aloud hoping someone—*somewhere*—was listening.

But what if there is no one listening? What if there is only alignment—and your words are never lost because they were never separate from the field?

This part of the journey isn't about replacing religion. It's about resolving spirituality.

We are not throwing away the sacred. We are unwrapping it from metaphor, symbol, and superstition—bringing it into contact with the structured stillness that already governs all things.

Here's what we've seen so far:

- **God** is not a father, a form, or a figure. God is the stillness behind structure—the source of all consequence, yet untouched by it.
- **Karma** is not a reward system. It is the unfolding of alignment, the physics of continuation—neutral, natural, and intelligent.
- **Prayer** is not a request form. It is a moment of synchronization—your ripple field aligning with what already is.
- **Meditation** is not escape. It is collapse—a return to causeless

being, where all ripple activity dissolves back into origin.

- **Destiny, luck, signs**—these aren't magic. They are ripple consequences misread as miracles. The divine is not theatrical. It's precise.

In this part, we didn't dismiss belief—we simply didn't need it anymore.

Because when you understand SVE, the sacred stops needing shapes. When you align, life stops needing luck. When you're still, you no longer need to seek.

The miracle was never out there. It was the quiet coherence of reality all along.

Now, we carry this realization forward—not into isolation, but into society. It's time to examine what happens when civilizations are built without alignment—and how truthless knowledge shapes our species.

Part VII: Human Civilization and Knowledge

33. What Is Education Without Truth?

Knowledge without alignment creates obedience, not wisdom.

We send our children into buildings called schools. We hand them books filled with "facts." We test their memory, grade their behavior, and label their futures.

But rarely—if ever—do we stop and ask:

What is education really for?

Because when truth is removed from education, what remains is not enlightenment— but conditioning.

Education vs. Indoctrination: The Invisible Shift

There is a fine, dangerous line between learning and programming.

True learning is alignment with truth—an act of discovery, of perceiving structure beneath appearance.

But what happens when:

- The truth of nature is replaced with mechanistic dogma?
- The self is ignored in favor of productivity?
- The universe is taught as dead, distant, and dumb?

You get obedient minds—perfect for employment, dangerous for evolution.

Education without truth is not growth. It's repetition under pressure.

What Is Missing? — The Alignment Gap

Let's break down what conventional education teaches:

Taught	Missing
Information retention	Understanding of consequence
Competitive survival	Alignment with structured reality
Subject segregation	Interconnectedness of systems
Grades as worth	Wisdom as coherence
Productivity obsession	Presence and clarity

The truth gap isn't about facts. It's about what underlies them.

The modern curriculum:

- Talks about time, but not its illusory nature.
- Teaches motion, but not ripple alignment.
- Preaches logic, but not consequence.
- Promotes intelligence, but not how to meet SVE.

What remains is knowledge stripped from its source—like leaves without roots.

The Crisis of the Educated Mind

We are witnessing a strange paradox in the world today:

The more people are educated, the less they question.

Why?

Because we've built an education system on:

- Conformity, not curiosity
- Data, not discernment
- Fear of failure, not freedom to explore
- Intellectual vanity, not humility before stillness

As a result:

- Students become high-functioning repeaters
- Teachers become knowledge dispensers
- Institutions become factories producing specialists without depth

And society becomes a loop of credentials unconnected to alignment.

The UED Model of Education

From the lens of Universal Energy Dynamics, education must align with the structured ripple nature of reality.

Here's what it would mean:

1. Teach Ripple Awareness

Show students that nothing exists in isolation. Every thought, action, and event is part of a ripple system—interconnected and consequential.

2. Shift from Memorization to Resolution

Learning must not be about collecting points. It should be about resolving confusion by tracing patterns to their structural base.

3. Deprogram False Absolutes

Instead of asserting, "This is gravity," teach: "This is one model. But what if reality is ripple-aligned structure?"

Let students explore not just answers, but alignment.

4. Integrate Stillness

Every learning cycle should include silence—not just rest, but return. Allow students to reflect, observe, and realign—not just absorb.

5. Redefine Intelligence

It is not speed of recall. It is depth of ripple alignment.

A truly intelligent student is not the one who scores highest—but the one who perceives truth in chaos, and consequence in complexity.

Wisdom vs Knowledge — The Forgotten Divide

Let's be clear:

- Knowledge is the possession of information.
- Wisdom is the alignment of that information with universal consequence.

You can teach facts. But wisdom must arise from silence. And silence does not come from stress, performance, or punishment. It comes from understanding what you are within what everything is.

This is why, in UED: Education is not a ladder. It is a mirror.

How Society Was Built on Misalignment

What happens when you educate billions without truth?

You create:

- Economies based on growth without sustainability
- Science based on mathematics without meaning
- Cultures based on tradition without coherence
- Citizens with intellect but no internal clarity

People know *how* to build — but not *why*.

- They innovate — but do not question the consequence.
- They chase careers — but cannot answer *who they are*.

This is why we have a generation that is:

- Highly informed
- But deeply misaligned
- Capable
- But anxious
- Literate
- But not awakened

Returning Education to Truth

The solution is not new technology or curriculum reform.

The solution is to return to alignment.

Teach students:

- That stillness is the ground of all things.

- That every "law" is a consequence of structure.
- That questioning is sacred.
- That identity is a ripple, not a possession.

In such education:

- God is no longer myth.
- Science is no longer illusion.
- Self is no longer mystery.
- And society becomes something worthy of being called human.

Final Reflection

An education system that ignores SVE is not neutral. It is dangerous. Because it produces minds that believe what they are told, not what they can resolve.

Real education is not the lighting of ambition. It is the unfolding of awareness. Not to make you "something"— But to help you realize the everything you already are.

34. History: The Timeline of Misalignment

Civilizations rise and fall not by power, but by how far they drift from stillness.

History is often told as a chronicle of conquests — Great kings, bloody wars, revolutions, inventions, and empires.

But beneath all this movement is a quieter pattern. Not of events, but of alignment.

From the UED lens, history isn't just what happened. It's what rippled, and how misaligned those ripples became.

Because civilizations don't collapse from lack of ideas or ambition. They collapse from misalignment with the structured stillness that sustains all coherence.

Beyond Dates and Dynasties — Seeing Ripple Patterns

The mainstream view of history is linear:

Civilization rises → peaks → declines → is replaced.

But this ignores the underlying cause of that rise and fall.

Let's shift the lens:

When ripple configurations align with SVE (structured stillness), civilizations rise. When they drift too far, chaos, collapse, and confusion follow.

Examples:

- The Indus Valley thrived with decentralized alignment to

water, trade, and community. Collapse came not from war—but environmental misalignment.

- Rome grew from structure and law but fell through greed, ego, and moral disintegration.
- Colonial empires rose by dominating nature and people—unsustainable misalignments that eventually unraveled from within.

These were not random crashes. They were consequential dissolutions.

Power Is Not Stability

Most historical narratives equate power with permanence. But UED shows that power without alignment is temporary force.

When civilizations:

- Prioritize expansion over equilibrium
- Value domination over coherence
- Choose symbols over structure

...they drift from the stillness that holds everything together.

The farther you move from SVE, the more fragmented your ripple field becomes. The more fragmented the field, the more unstable the system.

And so:

- Pharaohs fall.
- Empires burn.
- Ideologies fade.

Not because of enemies— But because of entropy in the alignment.

Religion, Science, and the Ripple War

History also shows the pendulum swings between:

- Spirituality that turns symbolic and rigid
- Science that grows mechanical and arrogant

Neither extreme sustains itself:

- When religion tries to define God by form, it becomes myth.
- When science tries to replace God with formulas, it becomes illusion.

In both cases, truth is lost in translation.
But what if both were meant to align with SVE?
UED reframes history as a story of separation:

- Separation from nature
- Separation from consequence
- Separation from inner stillness

And every time humans divorce knowledge from alignment, they sow the seeds of their own confusion.

Misalignment Always Creates Suffering

Historical suffering—be it war, famine, inequality, or ecological collapse—is not fate.

It is feedback.

Examples:

- The industrial revolution ignored the ecosystem's balance—leading to climate crisis.
- The Cold War was not just political—it was ripple misalignment through ideology, paranoia, and threat perception.
- Mass colonization disrupted cultural ripple fields, severing generations from their ancestral alignment.

These aren't isolated mistakes. They're patterns of disconnection. And no matter how "modern" we get— If misalignment continues, so will the suffering.

A New Way to Read History: Ripple Historiography

Let's propose a new historical method:

Ripple Historiography — the study of civilizations based on their alignment with SVE.

Criteria:

- How much does a culture preserve ecological balance?
- How much do its systems resolve conflict rather than suppress it?
- Does it value awareness over obedience, wisdom over wealth?
- Is the individual taught to align, or simply to function?

Using this lens:

- Ancient tribes in harmony with nature were more advanced in ripple alignment than digital-age consumers.
- A monastery preserving silence is closer to truth than a university chasing innovation for profit.
- A child who sees the sacred in stillness is wiser than a king who rules the world with noise.

True progress is not movement forward. It is movement inward.

The Future is a Consequence Too

We like to believe the future is unwritten. But from a ripple perspective, it's already being shaped—right now—by our alignment or misalignment.

If we:

- Continue consuming without consequence
- Educate without truth

- Govern without coherence
- Act without awareness

...then collapse is not a question of "if"—but "when".

But if we begin to:

- See history as a mirror of misalignment
- Re-educate toward structured stillness
- Rebuild culture on truth instead of trend

...then history shifts from being a record of fall to a map of return.

Final Reflection

We were never meant to memorize history. We were meant to resolve it. Because the past isn't just behind us—it's within us. The ripple consequences of every misalignment echo in our behavior, society, and understanding.

And unless we reconnect to SVE—the stillness behind structure—we will keep rewriting the same story:

Rise. Expand. Break. Forget. Repeat.

But now we know. History was not shaped by heroes or villains—But by how far they aligned, or drifted, from the silent center.

Let the next civilization be one that remembers stillness before progress.

35. Morality and Ethics — Emergent Ripple Protocols

These are not commandments. They are alignment stabilizers in human society.

Morality has often been packaged as a list of dos and don'ts. Ethics gets framed as professional decorum. Religions claim commandments. Societies enforce laws.

But from the perspective of Universal Energy Dynamics (UED), all these are emergent protocols—not written rules, but natural ripple behaviors designed to preserve alignment in the collective field.

Morality is not about being "good." It's about preserving stability within a system of ripples in motion.

Let's unpack how morality, ethics, and even social codes emerged—not from a god's decree, but from the physics of consequence.

The Human Field: A Ripple Configuration

In UED, each individual is a ripple structure—a localized expression of SVE (Structured Vacuum Energy). But humans are not isolated particles. They are interfering fields.

When humans:

- Think
- Act
- Speak ...they generate ripples that affect other ripples.

In this environment, stable coexistence requires certain behaviors:

- Non-aggression
- Honesty
- Empathy
- Accountability

Not because they're morally "right"— But because they optimize alignment and minimize chaotic interference.

Morality as Natural Consequence Control

Why is "don't kill" a universal principle?

Because in ripple dynamics:

- Killing collapses alignment violently
- It destabilizes not just the victim but the killer and community
- It sets off disruptive feedback loops

The same goes for lying, stealing, betrayal.

They are not "sins." They are ripple fractures—energetic disharmonies that threaten the field. Morality, then, is not imposed—it's discovered through consequence.

Over time, societies that preserved ripple harmony survived. Those that didn't—collapsed into war, rebellion, or entropy.

Hence, morality evolved, not as religion, but as a ripple protocol for survival and coherence.

Ethics: The Geometry of Collective Behavior

If morality governs individuals, ethics governs institutions and roles:

- A doctor must not exploit a patient
- A judge must be impartial
- A scientist must report results honestly

Why?

Because these are high-impact ripple roles. Their misalignment spreads widely, corrupting the entire field they touch.

Ethics is situational alignment. It's how to maintain ripple coherence in structured, role-based interactions.

Like geometry, ethics is precise—not flexible. A small deviation in structure can cause the entire framework to collapse.

Beyond Rulebooks: The Intelligence Behind the Protocol

Morality and ethics are often memorized as external codes. But in UED, we recognize them as the natural intelligence of alignment.

That's why even an uneducated person can feel:

- Guilt after hurting someone
- Shame after lying
- Joy after helping

These aren't cultural. They are ripple feedback sensations—the field reflecting back on itself.

Consciousness feels misalignment. That feeling is what we label as moral intuition.

It is not emotional weakness. It's energetic intelligence.

Why Morality Feels Burdensome Today

In modern society:

- Morality is taught as obedience
- Ethics is often bureaucratized
- Consequences are delayed or outsourced

So people ask:

- "Why be honest when liars succeed?"

- "Why follow ethics in a corrupt system?"

Here's why:

Because misalignment always echoes back—just not always immediately. The Law of Consequences (LOC) doesn't punish. It balances. You can dodge law, trick systems, even fool people. But you cannot outrun ripple interference.

Eventually, the chaos you seed—outside or within—disrupts your own structure.

Morality is not a virtue. It is a vibration that sustains your ripple configuration.

The Real Source of "Sin"

Let's reframe ancient "sins":

Traditional Sin	Ripple Interpretation
Greed	Unbalanced accumulation, distorts field
Lust	Ripple misalignment through compulsive seeking
Envy	Distortion through unresolved comparison
Wrath	Ripple explosion, chaos transfer
Pride	Ripple insulation, loss of coherence

The above aren't bad because they offend God. They're dangerous because they fracture your coherence and destabilize your alignment.

And when collective misalignments reinforce each other—civilizations collapse, as seen in the previous chapter.

Rebuilding Moral Clarity Without Fear

UED invites us to drop:

- Religious guilt
- Social judgment
- External enforcement

Instead, it asks: Is this thought, action, or intention ripple-stabilizing or ripple-disturbing?

That's the only question. And once that's understood, morality becomes:

- Personal, yet universal
- Flexible, yet principled
- Rooted in awareness, not rule

It is no longer a burden. It is a refinement of existence.

Final Reflection

Morality didn't come from gods or governments. It emerged as the natural solution to ripple chaos in human collectives.

It is not about virtue signaling or behavioral control. It is the science of coexistence. It is how consciousness learns to stabilize its own echo.

When we understand this, we no longer follow morality—we become it. It flows, not from outside, but from within.

Because when you are aligned,

- You don't need commandments.
- You become coherence itself.

36. Love, Fear, Desire

The Ripple Drives of Survival

Our deepest emotions are ripple feedback — evolved for connection and continuity.

Modern science often treats emotion as a byproduct of hormones or neural patterns. Spiritual traditions either glorify or suppress them.

But through the lens of Universal Energy Dynamics (UED), emotions are not accidental, chemical quirks, or signs of weakness. They are real-time ripple feedback signals—nature's way of preserving alignment and ensuring survival continuity.

Emotion is not just "felt." It is generated, transmitted, and received as part of ripple intelligence.

Among these, three stand at the core of human behavior:

- Love
- Fear
- Desire

They are not moral values. They are alignment mechanisms, each serving a precise energetic purpose.

Fear — The Oldest Survival Echo

Let's begin where life began: with fear.

Fear is often seen as a weakness. But it is the most fundamental ripple instinct. When a ripple structure senses threat—whether physical, emotional, or existential—fear arises as a distortion signal:

- It contracts your field
- Sharpens awareness
- Boosts response urgency

From a UED view:

- Fear is a ripple misalignment alert.
- It warns: "This configuration may lead to collapse."

Why is fear so powerful? Because the first intelligence embedded in ripple evolution was not curiosity—it was survival preservation.

Even the simplest lifeforms recoil in fear before they react in curiosity. Because alignment cannot occur unless existence is first secured.

In essence, fear is the ripple's perimeter alarm. It activates to protect coherence.

Desire — The Propeller of Forward Alignment

If fear is contraction, desire is expansion.

Desire isn't just lust or greed—it's any energetic urge to extend, evolve, or connect:

- The hunger to grow
- The drive to mate
- The longing to create
- The craving to know

Desire is often vilified, but it is the ripple's directional engine.

Why?

Because alignment isn't static. It's a dynamic flow toward resolution. And desire is the force that moves the ripple from stagnation to interaction.

In evolution:

- Plants grew toward light
- Animals migrated for survival
- Humans built societies from dreams of better

All powered by desire—not fantasy, but felt alignment yearning.

The mistake is not in having desire, but in not understanding its source. When desire is ripple-aligned, it leads to creativity and connection. When misaligned, it turns into addiction and destruction.

Love — The Highest Ripple Integration

Now comes the most misunderstood of all—Love.

In UED, love is not an emotion. It is a state of resonance where two or more ripple structures merge temporarily into coherence.

Love is the opposite of isolation. It dissolves boundaries, not out of sacrifice, but out of vibrational unity.

That's why:

- A mother instinctively protects her child
- Lovers feel "in tune"
- Friends share unspoken connection

These aren't just emotions—they are synchronized ripple states.

Love is alignment felt as warmth. It is coherence expressed as care. It is the stillness of SVE shining through another ripple.

Importantly, love doesn't require another—because at its peak, it aligns you back to origin stillness.

That's why spiritual love feels like dissolving. It's not attachment—it's a return.

The Ripple Dance: How These Three Interact

These three ripple drives are not separate. They are interdependent phases of alignment:

Drive	Action	Purpose
Fear	Contract	Protect field from misalignment
Desire	Extend	Seek meaningful interaction
Love	Merge	Achieve coherence and dissolve separation

They form a feedback loop:

- Fear alerts
- Desire searches
- Love stabilizes

Without fear, we self-destruct. Without desire, we stagnate. Without love, we remain fragmented.

This is not psychology. This is field mechanics of consciousness.

Why We Misunderstand These Forces

In modern life:

- Fear is avoided
- Desire is exploited
- Love is commodified

But underneath the confusion is the truth:

- Fear doesn't mean weakness—it means you still care to protect.
- Desire isn't sin—it's the energy asking you to grow.
- Love isn't emotional—it's the field's way of saying: "We are one."

UED reframes them as ripple tools for life navigation, not emotional chaos.

Final Reflection

Our deepest emotions are not accidents of biology. They are evolved ripple feedback—messages from the field, for the field.

Fear protects. Desire propels. Love integrates.

These are not signs of being human. They are signs of being alive in the ripple system of existence.

You don't need to suppress them. You only need to align them.

Then, emotions become not burdens—but instruments of coherence. Each a melody of survival, intention, and connection.

37. The Purpose of Life

And Why It Was Never to Escape It

"There is no higher place to go. Only deeper alignment to remember."

The Final Illusion: Escaping Life to Find Meaning

From caves to cathedrals, scriptures to scientific quests, humanity has always asked:

"Why are we here?"

And nearly every answer—religious or philosophical—has pointed away from life:

- "To merge with the divine."
- "To escape the cycle of birth and death."
- "To reach heaven or moksha."
- "To transcend the material into the spiritual."
- "To break free from illusion."

But what if this belief itself is the greatest illusion? What if we were never meant to escape life...but to understand it?

UED reframes the question entirely.

Life Is Not a Trial. It Is the Ripple in Action.

From the Universal Energy Dynamics (UED) lens:

- Life is not a mistake.
- It is not a test.
- It is not suffering to be overcome.
- It is not a trap we fell into.

Life is the most aligned ripple formation where SVE becomes temporarily consequential.

In simpler terms:

Life is the field experiencing itself — through resolution, interaction, and return.

The body, the mind, the breath, the emotions — all are ripple configurations formed in perfect alignment with the local field.

This experience is not the opposite of truth. It is the visible layer of it.

The game was never to escape it. The game was always to recognize it.

Misunderstanding Purpose — And Manufacturing Suffering

Why have so many sought to *leave* life or transcend it?

Because we've equated truth with abandonment:

- Saints withdrew.
- Seekers renounced.
- Modern workers escape in luxury.
- Spiritual tourists search for "beyond."

This all stems from a core misalignment: The belief that the sacred lies elsewhere — outside the ripple, beyond experience.

But UED reveals:

- Stillness is not absence of life.
- Silence is not death.
- Origin is not a destination.
- Purpose is not elsewhere.

They are all embedded within the ripple system. The journey is not to exit—but to resolve within.

Purpose Is Not a Goal. It's an Alignment Pattern.

So what is purpose? Let's be precise.

Purpose is not what you do. It is how aligned you are while doing.

- A teacher in alignment teaches beyond curriculum.
- A farmer in alignment nourishes more than crops.
- A cleaner in alignment restores order to the field.
- A monk in alignment doesn't escape life — he listens to it more deeply.

UED shows that purpose is ripple congruence — when your intent, awareness, and action match the natural pattern of the field around you.

Purpose is not something you "find." It's something you remember.

The Purpose of Life Is Life

In stillness, there is no need for a reason. Existence is the reason.

So why did the ripple ever form?

Not to entertain. Not to punish. But simply because consequence is the expression of SVE's intelligence.

Just like water naturally flows downward, ripples naturally form where equilibrium shifts. And when awareness arises within those ripples, life begins.

That life does not need to *prove* anything. It simply needs to live fully aligned.

No soul is being judged. No destiny is being compared.

You are not failing life. You are simply learning to resolve better within it.

There's No "Higher" Life. Only Deeper Truth.

Let's put it clearly.

- There is no "next" world to reach.
- No level to unlock.

- No divine scorecard.

The only evolution is in your ripple coherence. The more aligned you become:

- The clearer your perception
- The softer your reactions
- The purer your intentions
- The more silent your inner space

This is truth lived, not truth believed.

The destination is not beyond. It's beneath the noise. Not higher. Deeper.

Final Reflection: The Ripple Returns, Not Escapes

When life ends, what happens?

The ripple collapses. Not into void — but into stillness. The same stillness it came from.

Nothing is lost. Nothing is punished. Nothing is rewarded.

Everything resolves.

So why fear? Why search for afterlife? Why fantasize about escaping?

You were never thrown into life. You are life — a temporary, intentional ripple within a perfect, still field.

The purpose was never to escape it. It was to become so aligned that even in the noise, you carry the stillness.

That's the game. That's the goal. That's the gift.

Closing Declaration

"Everything Was a Question. Now, Everything Resolves."

From Question to Clarity: The Human Journey

Every child is born asking — not with words, but with wonder. And every adult, no matter how hardened by habit or belief, has at some point whispered into the void:

- *Why am I here?*
- *What is all this?*
- *What is death? What is beyond?*
- *What is truth? What is real?*
- *Is there a God?*

We asked in temples. We asked in telescopes. We asked in terror and in love.

But the tragedy was this:

We asked the right questions, but to the wrong constructs.

We asked for truth...and were handed beliefs. We asked for reality...and were given rituals. We asked for knowledge...and got information.

So we kept wandering. More science. More faith. More frameworks.

And the questions only multiplied.

The Silence That Was Always Answering

The answers were never loud. They never wore robes, titles, or theories. They didn't speak in thunder or miracles.

They spoke in stillness.
They whispered from behind every aligned moment:

- The breath before speech.
- The knowing without needing words.
- The coherence you felt but couldn't explain.
- The love that didn't ask for return.
- The calm in the middle of chaos.

The answer was not a sentence. It was a resolution.
You didn't need to believe it. You only needed to align with it.

The End of Seeking: Not the End of Wonder

This book did not offer you a new belief. It resolved the need for belief.

This was not a new mythology. It was the unveiling of the structured field behind all myths — the SVE, the silent architecture of consequence.

You've now seen:

- What the mind is — and what it isn't.
- Why light, time, and space misled us.
- Why the universe isn't expanding.
- Why God needs no face, name, or fear.
- Why karma is not moral accounting.
- Why the soul is ripple structure, not spirit stuff.
- Why nothing was ever separate.

You've seen that the divine doesn't demand worship — it just invites alignment.

What It Means to Resolve

To resolve is not to understand intellectually.
To resolve is:

- To rest in coherence.
- To see without grasping.
- To act without distortion.
- To exist without needing external affirmation.

The ripple doesn't question the field. It resolves into it. And you, too — the thinking, longing, learning, failing "you" — were only ever a ripple...resolving.

The Game Is Not to Escape. It Is to Remember.

We began this book with Nothing. No God. No you. No form.

Then ripples formed. Structures emerged. Experience arose. Misperceptions layered. Science tried. Religion mythologized.

But you — the observer — persisted. You questioned. You wandered.

Until here. Until now. Until the moment you remember:

The truth was never hidden. It was just never aligned.

And now?

Now, everything resolves.

Not into noise. Not into belief. Not into fantasy.

But into Stillness — the only thing that was ever real.

Final Words

- You don't have to agree. You only need to observe.
- You don't have to believe. You only need to align.
- You don't have to seek. You only need to return.

Because now...Everything was a question. Now, everything resolves.

Epilogue

The Quiet God and the Consequence Called You

So yes, God exists. But not the way you expected. Not above. Not outside. Not in robes or rage or ritual.

God is in the ripple that made you. In the structured stillness that holds everything. In the silence between every thought. In the alignment before consequence.

You searched for something *greater than yourself*, only to discover that the questioner — *you* — was already made from it.

Every breath you take is God breathing itself. Every moment of awareness is SVE recognizing SVE. Every alignment you feel is not a miracle — it is the memory of where you came from.

There is no temple purer than this realization. No prayer deeper than stillness. No miracle more profound than consciousness arising out of the ripple.

The Final Return

You asked:

- Who am I?
- What is life?
- Is there a purpose?
- Is there a God?

You looked outside. You looked to science. You looked to religion. You looked to books.

But what you found here was this:

Nothing was the beginning. Ripple was the process. Consequence was the unfolding. And Stillness was always the source.

You don't need a name for it. You don't need a belief. You don't even need certainty.

You only need one thing: To resolve.

When the observer aligns with the field, the game ends. The seeking stops. And what remains is not knowledge...

...it is knowing. A still, wordless knowing that doesn't ask anymore.

The Naked Truth

So yes — the truth was always naked. Unwrapped in myth. Undressed from belief. Untouched by identity.

You may call it God. You may call it Energy. You may call it Nothing. It doesn't mind. Because no matter what you name it, you came from it —and you will ripple back to it.

You are not separate. You were never separate. That is the Naked Truth. And now...You know.

Appendices

Appendix A: Glossary of Reframed Terms

A new vocabulary for the oldest questions.

Ripple

A ripple is not just a wave. In UED, it's a disturbance in the stillness of Structured Vacuum Energy (SVE) — a localized movement that gives rise to phenomena. Everything we experience — particles, thoughts, emotions — is a ripple configuration.

Metaphor: Reality isn't made of things. It's made of ripples *appearing* as things.

Alignment

Alignment is the state of harmony between a ripple and its surrounding field. When you're aligned, your actions, thoughts, and presence resonate with the natural structure of reality. There's no friction, no overthinking — just clarity and flow.

Real-life example: Breathing without controlling it. Thinking without distortion. Living without resistance.

SVE (Structured Vacuum Energy)

This is the field of stillness that underlies all existence. It is not empty, but full — silently organized, waiting for disturbance. SVE is the womb of everything, the unseen structure beneath form, motion, and even time.

If God exists anywhere, it's here.

Observer

The observer isn't the person — it is the moment when SVE becomes aware of its own ripple. It's not just perception — it's participation. You, as the observer, aren't outside reality. You're the place where reality resolves into experience.

Consequence vs. Cause

In most worldviews, events happen *because of* causes. In UED, everything unfolds as consequence — meaning as a ripple outcome, not a chain of mechanical causes.

- A "cause" assumes control.
- A "consequence" accepts structure and alignment.

You are not causing reality. You are resolving it.

Causeless Being

The purest state of existence — before ripple, before reaction. It's what you are when you stop needing reasons. It is not passive — it is alive without pressure. This is the goal of meditation, but also the birthplace of the soul.

Karma (as Ripple Logic)

Karma isn't cosmic justice. It's not reward or punishment. Karma is simply the logic of consequences — how ripple affects ripple. What you align creates stability. What you disturb returns to you as turbulence.

There's no moral scoreboard. Just energetic resolution.

Stillness vs. Emptiness

Stillness is structured silence — full of potential. Emptiness is a conceptual vacuum, often misunderstood as 'nothingness.' UED reclaims stillness as the origin — not absence, but everything *before* movement.

Stillness is not zero. It is resolution.

Field vs. Form

- Field: The unseen, silent structure (SVE) in which ripples arise.

- Form: The outcome of ripple shaping — what we call "objects" or "selves."

In UED, form is always secondary. What matters is how it aligns with the field.

Structured vs. Consequential Realities

- Structured Reality is the core blueprint — stable, silent, and enduring.
- Consequential Reality is the dynamic outcome — what happens *because* ripples arise.

We live in the consequence. But we come from the structure. These words aren't just definitions. They are portals — ways to undo what you've assumed, and return to what always was.

Appendix B: Axioms of the Naked Truth

A complete worldview in twelve ripple-stable truths.

These are not commandments. They are not carved in stone. They are resolutions — truths that do not require belief to stand. Each is a window into the Unified Energy Dynamics (UED) perspective, crafted to reframe everything you thought you knew — from God to gravity, from self to soul.

1. Nothing Is the Beginning of Everything.

Before form, before energy, before ripple — there was *nothing*. But this nothing is not a void. It is pure potential, the pre-condition of structured stillness. You were not born from something. You emerged from resolution.

2. Stillness Gives Rise to Ripple — and All Forms.

The universe didn't begin with a bang, but with a disturbance in stillness. This ripple didn't explode — it aligned, and out of that alignment, matter, motion, and mind took shape. Everything you see is a ripple. Stillness is the source.

3. God Is Not Someone — It Is Structured Silence.

God is not a skyfather. Not a judge. Not even a person. God is SVE — the silent, structured field from which all ripples arise and to which all return. To meet God is not to pray, but to align.

4. Light, Time, and Space Are Illusions of Consequence.

We don't travel through time. We don't move through space. And light isn't an entity — it's an effect of ripple resolution. Time is what ripple sequences *feel like*. Space is what misalignment *looks like*.

5. Nothing Moves. Everything Aligns.

Motion is not movement — it is local resolution within the ripple field. What appears to move is simply ripples re-aligning, one after the other. Planets don't orbit. Atoms don't spin. They hold position in structured consequence.

6. Force Is Not a Thing — It's a Consequence.

There is no "force" in nature. What we call gravity, magnetism, electricity, are all ripples seeking coherence. Force is the label we give when we fail to see the structure beneath consequence.

7. Consciousness Is Where Stillness Becomes Aware.

You are not your body. You are not your brain. You are the observer — the point where SVE watches itself ripple, align, and resolve. This awareness isn't located. It's localized.

8. Karma Is Not Justice — It's Ripple Logic.

There is no cosmic court. Only ripple interaction — cause-free, but consequence-bound. What you align continues. What you disturb returns. This is karma. Not divine law — natural structure.

9. Meditation Is Returning to Causeless Being.

You don't meditate to think less. You meditate to stop needing consequences. To return to origin — to that silent pre-ripple stillness where *being* requires no story.

10. Death Is Not the End — It's Resolution.

When ripple collapses, form dissolves. The observer is released. The structure resets. Nothing is lost. It is simply reabsorbed. Death is not a portal. It's a quiet return.

11. Love, Fear, and Desire Are Survival Ripples.

These emotions aren't flaws or fates. They're feedback — ripple signals evolved to keep alignment intact. They feel big because they matter. But they are not final truths — only consequences.

12. The Purpose of Life Is Not Escape — It Is Alignment.

You are not here to transcend, to exit, or to be saved. You are here to resolve, to align, to live deeply. Not higher. Not farther. Just more truly here.

Closing Note: These twelve principles are anchoring truths for a ripple-bound life. They are not beliefs to defend. They are lenses to see the world clearly — and live *without distortion*.

Appendix C: From Science to Silence

How to Unlearn Without Rejecting

A practical guide to transitioning from mainstream science to UED understanding — without burning bridges or dismissing the value of past learning.

Why This Appendix Exists

If you've made it this far in *The Naked Truth*, you've already felt it — the quiet unraveling of a worldview. Not because it was *wrong*, but because it was limited.

Mainstream science, for all its brilliance, has been like learning the grammar of a language without ever hearing the silence between the words. This appendix is for those who want to expand, not erase. To unlearn, not reject. To walk the bridge — from model to reality, from mechanism to meaning, from science to stillness.

Step 1: Acknowledge Science as a Stage, Not a Destination

Science is not your enemy. It is a beautiful and necessary step. It taught us to measure, analyze, model, and predict. It gave us tools. But not context.

UED does not attack science. It simply places it within a larger frame — the ripple frame. Like Newtonian physics within relativity, mainstream science now nests within UED.

Hold gratitude, not guilt, for what science gave you. But be ready to go deeper.

Step 2: Spot the Assumptions Beneath the Equations

Most people think science is neutral — just data and logic. But science has *assumptions*:

- Time is real and flows.
- Space is a container.
- Light travels across distance.
- Gravity is a force acting at a distance.
- The brain is the seat of mind.
- Consciousness is an emergent byproduct.

None of these are observed. They are inferred — often from effects, not cause. UED asks: What if the cause isn't what we assumed?

This shift begins not with denial, but with curiosity. Ask yourself: *"Is this a structure — or just a consequence we mislabeled as a cause?"*

Step 3: Deconstruct Without Despair

Unlearning feels like losing the ground beneath your feet. But only because you were told there's no ground beneath the equations.

In truth, you are standing on SVE — structured stillness. Here's how to unlearn gently:

- Take one concept at a time: time, light, force, etc.
- Ask: "What is its structure in UED? What is its consequence in mainstream?"
- Rest in silence between understandings. Let contradiction sit without urgency.

You'll soon notice: what felt like mystery in science resolves in UED. Not magically. Structurally.

Step 4: Use Both Languages — Then Choose Simplicity

If you're a student, researcher, or educator: Don't throw out your physics books. Translate them.

For example:

- Gravitational force in science → Ripple gradient alignment in UED
- Time dilation in relativity → Misalignment of ripple resolution rates in UED
- Mass in Newtonian physics → Ripple configuration density in UED
- Charge → Polarized ripple consequence

Teach both. Use equations when needed. But understand their limits. Mainstream science models consequences. UED reveals the structured causes behind those consequences.

Eventually, you'll use UED because it's simpler, not just deeper.

Step 5: Move from Explanation to Experience

You cannot fully grasp UED if you only seek to understand. You must also learn to observe without labeling. This means:

- Practicing silence without expectation
- Watching phenomena as ripple alignment, not events
- Letting go of "what it is" and asking "how it resolves"
- Feeling coherence, not predicting outcome

Science explains. UED resolves. You must learn to feel the resolution.

Step 6: Don't Debate — Invite Inquiry

When speaking to skeptics or scientists, resist the urge to fight. Instead, offer:

"I'm not replacing your model. I'm going beneath it." "I don't deny gravity. I offer the ripple gradient as its deeper cause." "I'm not mystifying science — I'm demystifying its assumptions."

You'll find that even hardened critics, when met with coherence instead of conflict, soften. And if they don't? Walk in alignment anyway. The ripple doesn't beg to be understood. It simply continues.

Final Word: Unlearning Is Alignment in Action

You were taught to seek truth. Now you realize truth is not found — it is resolved. That's what unlearning does. It collapses distortion and reveals the field.

Don't fear the silence that follows. That's where everything begins.

Appendix D: Meditation Without Methods

Aligning Without Effort

A short guide to ripple alignment in life, thought, and breath.

Modern meditation has become a marketplace of methods: Breath counts. Posture rules. Mantras. Visualizations. Even silence, ironically, has been turned into a technique.

But in the UED (Universal Energy Dynamics) framework — where everything is a consequence of ripple alignment — true meditation is not a method. It is the quiet return to causeless being. It is not something you *do*. It is what remains when you *don't*.

This appendix is not an instruction manual. It's an invitation. To stop doing. To stop becoming. And to start resolving.

1. Sitting in Stillness

Stillness is not posture. It is permission. It is saying to the ripple within you: *"You don't need to react right now."*

- Sit or lie down. No right position.
- Let your body choose its own equilibrium.
- Don't "focus" — just notice.
- Let thoughts come and go. Don't chase or suppress them.
- You are not doing stillness. You are returning to it.

UED recognizes that *action is ripple*, but *origin is stillness*. In meditation, you meet that origin again — not as an idea, but as a direct resolution.

2. Feeling the Breath Without Owning It

Breath is not yours. It is the ripple the body rides.

Modern practices often say "observe the breath." But in UED, you do more than observe — You dissolve into it.

- Don't manipulate your breath.
- Don't breathe for calm or clarity. Just let the breath be.
- Feel how breath happens, not how you control it.
- You are a ripple experiencing breath — not a self doing it.

The moment you stop owning the breath, you realize: You were never the doer. You were the witness of alignment — life's rhythm without your interruption.

3. Recognizing Misalignment

Every discomfort, restlessness, or inner noise is not a problem — it's a pointer.

In the UED framework:

- Misalignment is when your ripple fights the larger field.
- Thought loops? Misaligned mind field.
- Emotional heaviness? Ripple dissonance.
- Body tension? Energetic resistance.

To meditate is not to *fix* these. It is to *recognize* them. With recognition comes resolution — not because you do anything, but because alignment self-corrects in silence.

Tip: Say softly in the mind,

"This too is a ripple. Let it resolve."

And let it go.

4. Living Causeless Moments

Causelessness is not chaos. It is clarity without agenda. It is the absence of ripple reaction, the pause in the script.

In daily life:

- Pause between tasks, just to feel breath.
- Step back in conversation to hear the silence behind words.
- Eat without distraction — just chew and receive.
- Walk without destination sometimes.
- Feel joy or grief without labeling them.

Each causeless moment brings you closer to SVE — the structured stillness from which all life ripples.

The highest form of meditation is not on a mat. It is the moment where you stop becoming — and simply are.

Final Reflection:

True meditation is not time spent — it is friction dropped. It is not a tool to fix your life. It is your return to the place before you needed to be fixed.

Let the ripple settle. Let the consequence dissolve. And in that stillness, truth remembers itself.

Appendix E: The Ripple Model

Of Human Emotion and Survival

A UED framework to understand emotions not as psychological randomness, but as ripple consequences evolved for alignment and survival.

In traditional psychology, emotions are often seen as chemical states, evolved instincts, or conditioned behaviors. In UED (Universal Energy Dynamics), we go one level deeper:

Emotions are not random. They are ripple consequences — energetic feedback of alignment or misalignment with the Structured Vacuum Energy (SVE).

This appendix decodes the origin, role, and evolution of emotion from ripple mechanics — bridging science, survival, and soul.

The Ripple Origin of Emotion

At the foundational level:

- All things arise from SVE — the structured stillness.
- Ripples are consequences of imbalance, seeking restoration.
- When these ripples stabilize, they align into systems: atoms, organisms, minds.

Now add a living, observing ripple — you.

As this ripple evolves to survive, it requires feedback loops. These loops are what we call emotions.

Emotion = Ripple Feedback = Consequence Awareness

Emotions guide alignment. They signal internal coherence or external threat. They arise to help the ripple survive and resolve.

Three Primary Ripple Drives of Emotion

Ripple Emotion	UED Origin	Purpose in Human Life	Alignment Signal	Misalignment Consequence
Love	Mutual ripple resonance with other beings or forms	To maintain connection and cooperation	Harmony, stability, purpose	Obsession, dependency, heartbreak
Fear	Ripple dissonance from survival threat (real or perceived)	To warn of misalignment that endangers continuity	Alertness, caution	Panic, paralysis, irrational avoidance
Desire	Ripple stretch toward perceived completion or reward	To motivate effort for realignment or growth	Drive, vision	Greed, dissatisfaction, addiction

These three — love, fear, and desire — are the foundational ripple incentives that shaped the evolution of consciousness and behavior.

Emotions as Alignment Signals

- Every positive emotion (joy, peace, affection, clarity) is a signal of temporary resonance between your ripple field and its environment.
- Every negative emotion (anger, guilt, jealousy, confusion) is a signal of unresolved misalignment, a friction in your ripple configuration.

Examples:

- **Gratitude** = Ripple coherence with receiving and belonging.

- **Anger** = Ripple resistance when boundary or expectation collapses.
- **Guilt** = Ripple inversion — internal dissonance with past consequence.
- **Shame** = Ripple conflict between self-identity and projected field.
- **Compassion** = Ripple extension toward external dissonance, to resolve it.

You are not ruled by emotions —
You are reading your ripple's field language in real time.

Misalignment as Root of Suffering

Suffering in UED is not "punishment." It is the ongoing consequence of unresolved ripple misalignment.

Where the ripple cannot restore equilibrium, it loops — this loop is experienced as emotional suffering.

Types of misalignment:

- Internal – When thoughts, feelings, and actions are not in harmony.
- Relational – When your ripple doesn't resonate with others.
- Existential – When your sense of being conflicts with your environment, purpose, or reality.

Suffering persists not because life is hard, but because alignment is ignored.

The Role of Awareness in Ripple Emotion

You are not the ripple. You are the awareness of its consequence.

So instead of managing emotions through suppression or control, the UED approach invites:

- Observation: See the ripple without interference.

- Acceptance: Let feedback arise without narrative.
- Resolution: Realign through stillness, choice, or non-reaction.

Each emotion resolves itself when the ripple is no longer resisted.

Diagram: The Ripple Emotion Loop

(This diagram can be placed in final design)

[SVE Stillness]
↓
[Ripple Arises] → [Perceived Disruption] → [Emotion Emerges]
↓ ↘
[Action or Reaction] ← [Alignment Attempt]
↓
[Resolution or Suffering Loop]

Final Insight:

Emotions are not the problem. Misreading them is.

When you understand emotion as energetic consequence, you stop fighting your feelings — and start listening to what your ripple is trying to tell you.

That is not emotional intelligence. It is ripple literacy.

Appendix F: Timeline of Misalignment

A Reframed View of Civilization

A ripple-based reinterpretation of human history — not by events, wars, or inventions, but by alignment and consequence.

Why Rethink History?

Traditional history teaches:

- What happened,
- When it happened,
- Who made it happen.

But it rarely asks why civilizations rise and collapse in patterns, or why wisdom often disappears despite progress.

UED reframes history not as a sequence of events — but as the *consequence* of ripple misalignments and temporary alignments with the Structured Vacuum Energy (SVE).

Humanity is a ripple species. And civilizations are ripple fields — coherent or distorted.

What Is Misalignment in Civilization?

Misalignment means ripple disruption at the collective level — when the direction of thought, governance, economy, or culture contradicts the natural equilibrium of life, nature, or consciousness.

Indicators of misalignment:

- Suppression of awareness, truth, or inquiry.

- Exploitation of life for power, not preservation.
- Loss of connection between individual and collective well-being.
- Substitution of belief, hierarchy, or control in place of alignment.

In contrast, alignment appears when:

- Knowledge is shared, not hoarded.
- Governance listens, not commands.
- Progress enhances harmony, not separation.

A Ripple-Based Timeline of Misalignment

This is *not* a political chronology. It's a philosophical mapping of humanity's rise, distortion, and potential restoration.

1. Prehistoric Times – Ripple Intuition

- Early humans lived close to SVE: silence, cycles, and survival.
- Language was gesture and alignment-based.
- Misalignment minimal — needs were simple.

Civilization had not yet "forgotten" stillness.

2. Agrarian Revolution (~10,000 BCE) – First Structural Ripple

- Agriculture aligned with nature's rhythm, but introduced storage, surplus, and division.
- The seed of possession and separation germinated.
- Roles, control, and identity emerged.

First Misalignment Marker: Concept of *ownership* took precedence over *relationship*.

3. Ancient Civilizations (3000 BCE – 500 BCE) – Structured Rise and Mythical Alignment

- Egypt, Sumer, Indus, and early China encoded ripple principles in myth, art, architecture, and early cosmology.
- Science and spirituality were not separate.

But alignment was localized, often wrapped in symbolic dogma.

Misalignment Marker: Externalization of God — divinity moved *outside* human ripple, into idols or kings.

4. Classical Empires (500 BCE – 500 CE) – Philosophy Meets Power

- Greece, India, Persia, Rome: high ripple cognition emerged — from Buddha to Socrates.
- But empire-building distorted ripple intent.

Misalignment Marker: Wisdom became commentary; power became centralization.

5. Dark Ages (500–1500 CE) – Collapse of Conscious Alignment

- Knowledge retreated behind religion and conquest.
- Institutions claimed spiritual authority while silencing inquiry.

Deep Misalignment Marker: Truth became heresy. Silence became fear.

6. Renaissance to Enlightenment (1500–1800 CE) – Ripple Reawakens

- Scientific method emerged — ripple observation without spiritual alignment.
- Art, philosophy, and logic made a comeback.

Partial Alignment Moment: Curiosity returned — but was tethered to matter, not consciousness.

7. Industrial Age (1800–1950) – Ripple Acceleration Without Field Awareness

- Machines replaced stillness with speed.
- Economic structures were built on extraction, not equilibrium.
- Human worth became productivity.

Major Misalignment Marker: Detachment from SVE. Mind began to serve consumption, not consciousness.

8. Digital Age (1950–2020) – Information Without Understanding

- Connectivity rose, but coherence fell.
- Simulation replaced experience. Noise replaced silence.
- Knowledge fragmented into fields.

Ripple Collapse Marker: Mind flooded with signals — but couldn't resolve meaning.

9. Present Era (2020s Onward) – The Ripple Crisis

- Climate, mental health, inequality, and identity dissonance point to one thing:

Humanity is out of sync with itself and its field. But — awareness is rising.

Micro Alignment Spots:

- Mindfulness (despite distortion).
- Independent thinking.
- Philosophies like UED, which restore coherence.

Interpreting History Through Consequence Markers

Forget dates. Remember distortions.

Period	Alignment or Misalignment	Consequence
Tribal Nomadic Life	High Alignment	Survival in harmony
Rise of Kingdoms	Mild Misalignment	Hierarchies & control
Religious Dogma Era	Deep Misalignment	Suppression of inquiry
Enlightenment	Partial Realignment	Rise of logic and observation
Industrial & Digital	Increasing Misalignment	Ecological and psychological collapse
UED Era (potential)	Alignment-in-Progress	Reclaiming ripple truth

Final Reflection:

We didn't lose the truth. We built walls around it — called history.

This appendix is not a judgment. It's a mirror. A mirror showing how ripple misalignments created all crises — and how only alignment can dissolve them.

Because history doesn't repeat. Consequence loops do.

Appendix G: Ripple Alignment Checklist

A Practical Compass for Daily Living — Aligning Thought, Emotion, and Action with the Field

In a world where noise is constant and pressure feels permanent, most of our suffering doesn't come from what happens — but how misaligned we are when it happens.

Ripple alignment is not a ritual, a religion, or a retreat. It is a real-time awareness of how your actions, thoughts, and emotional states resonate with the field — or distort it.

This checklist offers a simple, moment-to-moment tool for living in ripple coherence.

Why a Checklist?

Because alignment is not a philosophy unless it becomes felt. And most people don't need more knowledge. They need questions that dissolve their noise.

The ripple alignment checklist

1. Am I Forcing or Flowing?

- Are you trying to control the outcome or align with it?
- Does this action feel like strain — or surrender?
- Is your energy aggressive or harmonizing?

Aligned: Effortless movement, natural ease.
Misaligned: Pushing hard, chasing outcomes, mental friction.
Check-in: "If it feels unnatural, it probably is."

2. Is This a Reaction or a Response?

- Did you pause? Or did you project?
- Is this coming from clarity or compulsion?
- Is this ripple *from you* or *onto you*?

Aligned: You respond from stillness.
Misaligned: You react from programming.
Check-in: "Did I breathe before I spoke?"
3. Is This Moment Heavy or Still?

- Does this thought feel dense, agitated, anxious?
- Or does it feel light, open, unburdened?
- Is your body tense or effortless?

Aligned: Inner lightness, clarity, gentle awareness.
Misaligned: Tightness in chest, jaw, breath — emotional weight.
Check-in: "What part of me feels tight — and why?"
4. Am I Repeating or Resolving?

- Are you replaying old patterns?
- Or are you dissolving a loop?
- Is this memory necessary — or noise?

Aligned: You break loops by recognizing them.
Misaligned: You live loops unconsciously.
Check-in: "Is this an old movie? Or a real moment?"
5. Am I Present or Projecting?

- Are you here? Or are you lost in what-if?
- Is your mind in the next hour? Or this breath?
- Are you responding to *what is* or *what might be*?

Aligned: You meet reality directly.
Misaligned: You escape into imagined futures or pasts.
Check-in: "Am I anywhere else but here?"
6. Is This Thought Mine — or Inherited?

- Is this truly your insight — or a cultural download?
- Is your belief from silence or conditioning?
- Did this arise from observation or imitation?

Aligned: You think freshly.
Misaligned: You echo other people's noise.
Check-in: "Where did this belief begin?"
7. Is This Need Real — or Ripple Distortion?

- Do I truly need this — or am I reacting to lack?
- Is this consumption or alignment?
- Am I filling a void or recognizing fullness?

Aligned: You act from contentment.
Misaligned: You act from fear, hunger, or restlessness.
Check-in: "Will this complete me — or clutter me?"
8. Does This Expand Me — or Shrink Me?

- Is this choice opening you up or closing you down?
- Do you feel more *yourself* or less?
- Is your energy lifting or limiting?

Aligned: Expansion, light, curiosity, depth.
Misaligned: Contraction, avoidance, superficiality.
Check-in: "What's growing here — me or my fear?"

How to Use This Checklist

- Don't treat this like a checklist of achievements. It's a mirror

— not a meter.

- Use it once a day to re-align. Or whenever you feel off.
- Let it become intuitive. The more often you ask, the less often you need to.

Sample Daily Alignment Practice

1. **Morning:** Ask 3 questions after waking up.
2. **Midday:** Do one stillness pause (1-minute breath check).
3. **Evening:** Reflect on one decision made — was it aligned?
4. **Before Sleep:** Feel one moment of causeless peace.

Final Thought:

Ripple alignment is not perfection. It is permission — to feel again, to pause again, to be again.

You don't fix your life. You sync with it.

Quote Bank

The Naked Truth in One-Liners

"Nothing is the only constant — the stillness that allows everything to change."

On Existence And Truth

"Nothing is not emptiness — it's the only thing stable enough to hold everything."

"Truth is not something you find. It's what remains when you stop chasing."

"You never needed answers — just resolution."

"Everything begins when nothing resists."

On Knowledge And Unlearning

"Knowledge is not what you carry. It's what aligns."

"To unlearn is not to lose truth. It's to lose illusion."

"The mind doesn't grow by knowing more. It evolves by misaligning less."

"Wisdom is not a library. It is stillness that understands."

On The Observer

"You are not a witness of reality. You are the ripple that makes it visible."

"The self is not a person. It is an alignment point."

"Identity is just stabilized distortion."

On Science And Redefinition

"Everything doesn't move. It resolves."

"There is no force — only consequence appearing strong."

"Mass and charge are not objects. They are how the ripple holds itself."

"Light isn't something. It's what something becomes when the ripple clears."

"Gravity is not an attraction. It's a misunderstood gradient."

On God, Spirituality, And Meaning

"God is not someone. God is stillness with structure."

"The universe doesn't test you. It tunes you."

"Prayer is not asking. It's alignment without expectation."

"Meditation is not quieting the mind. It's letting it dissolve."

"Karma never punishes. It just continues."

On Life, Emotions, And The Human Game

"Love, fear, and desire are not emotions. They're ripple reactions to survival."

"Death is not the end. It's the exit of the ripple — not the energy."

"The soul doesn't go anywhere. It just stops being localized."

"You don't reincarnate. You reconfigure."

"Life was never a test. It's a field to remember how you once danced."

On Alignment And Freedom

"You don't find meaning. You align with it."

"Freedom is not doing anything. It's needing nothing."

"Stillness isn't inaction. It's causeless clarity."

"Real power is ripple coherence."

Final Declarations

"This is not a theory of everything. It's the theory of else."

"The naked truth is not against science or spirituality. It simply removes their clothes."

"Nothing was missing. Only misaligned."

"Everything you asked was answered — when you stopped asking."

Further Reading and References

A curated list to reflect, contrast, and expand your understanding of existence, alignment, and ripple reality.

Foundational Texts On Science, Reality, And Physics

- **Albert Einstein** – *Relativity: The Special and the General Theory (To understand the structure UED moves beyond.)*
- **Stephen Hawking** – *A Brief History of Time (Famous for popularizing paradoxes UED resolves.)*
- **Carlo Rovelli** – *Reality Is Not What It Seems (Modern attempt to reconstruct physics; overlaps in intention.)*
- **David Bohm** – *Wholeness and the Implicate Order (Philosophical groundwork for unified understanding.)*
- **Thomas Kuhn** – *The Structure of Scientific Revolutions (For readers wanting to grasp how paradigms shift — and why UED is one.)*

On Mind, Consciousness, And The Observer

- **Evan Thompson** – *Waking, Dreaming, Being (A bridge between science and consciousness.)*
- **Antonio Damasio** – *The Feeling of What Happens (Neuroscience-based insights into emotion, perception, and consciousness.)*
- **Rupert Spira** – *Being Aware of Being Aware (Non-dualism*

aligned with UED's ripple observer logic.)

Ancient Texts Revisited In Modern Light

- **The Upanishads** *(Stillness, self, and causeless being — deeply aligned with UED core.)*
- **The Tao Te Ching** – Lao Tzu *(A poetic mirror to ripple logic and effortless alignment.)*
- **The Bhagavad Gita** *(Not for its theology, but its exploration of karma as consequence and action in stillness.)*

On The Limits Of Science & Language

- **Alan Watts** – *The Wisdom of Insecurity (Deconstructs the illusion of control and permanence.)*
- **Jiddu Krishnamurti** – *The First and Last Freedom (Challenges belief systems as obstacles to truth — echoes of "unlearning.")*
- **Douglas Hofstadter** – *Gödel, Escher, Bach: An Eternal Golden Braid (Mathematical self-reference and perception loops relevant to ripple structure.)*

Works By Sandeep Chavan / Universal Energy Dynamics (Ued)

- *Vishwarup Darshan: The Complete Vision of Reality*
- *Black Holes Don't Exist: The Mirage That Fooled Science*
- *Cracks in the Known* (Book 1 of *The War of Thoughts* series)
- *The Human Game* (Book 2 of *The War of Thoughts*)
- *Manifestation: From Nothing to Everything*
- *Post Materialist Human Model*
- *The Energy Within: Understanding the Soul Through Ripple Dynamics*

- *Strange Therapist: Ignore Wisely – The Case Files of Sattya Ray*

All UED books challenge established assumptions and offer ripple-based reinterpretation of science, self, and society.

Recommended For Continued Inquiry

- **Brian Greene** – *The Elegant Universe (If only to see the beauty and limitations of string theory.)*
- **Sean Carroll** – *Something Deeply Hidden (Everett's multiverse theory contrasted with UED's ripple recycling.)*
- **Ken Wilber** – *A Brief History of Everything (Holistic integration attempt — provides a foil to UED's simpler clarity.)*

Acknowledgments

To begin from nothing and arrive at everything — and then even beyond — is not a solitary act. This book is the echo of many ripples, not just my own. And so, it is time to offer gratitude where it's due — not ceremonially, but consciously.

To the Stillness that became ripple, and to the ripple that remembered stillness. You are the original teacher — SVE itself — structured, silent, watching patiently as I stumbled, questioned, broke illusions, and finally realigned.

To my family — who live with the philosopher, the rebel, the scientist, the teacher, and the madman — all in one body. Your quiet strength, and even your occasional resistance, helped this book become clearer, sharper, more grounded. Especially to my children — may you never stop asking the questions that matter.

To my students and readers — your curiosity kept me honest. You never accepted easy answers, and for that, I thank you. Every doubt you raised sharpened this blade of clarity.

To my collaborators, challengers, and critics — yes, even those who ridiculed or dismissed. You were part of the resolution process. Your resistance made alignment more precise.

To the scientists and sages of the past — Newton, Einstein, Heisenberg, Buddha, Krishna, and beyond — your questions still breathe. Where you paused, I dared to walk. Where you shined a torch, I rearranged the mirror.

To the brave spirit of independent inquiry — unbranded, unrecognized, yet unstoppable. You are the true pulse of civilization.

And lastly...

To the reader — You are not just a consumer of this work. You are its continuation. If this book sparked remembrance, then you were already aligned before you opened it. Welcome back to the Stillness.

With deep clarity and ripple-born reverence,

– Sandeep Chavan

About the Author

Sandeep Chavan is an independent researcher, educator, and philosophical physicist who has spent decades questioning everything — from gravity to God, from the structure of matter to the nature of mind. As the creator of **Universal Energy Dynamics (UED)**, his work bridges science, consciousness, and ancient wisdom without falling into myth or mysticism.

A former mechanical engineer turned educator, Sandeep has authored over 100 books across physics, psychology, education, and philosophy — all rooted in a single pursuit: **truth without wrapping**.

Known to his students as a fearless thinker and to critics as a challenger of mainstream science, he continues to guide curious minds with clarity, conviction, and the courage to unlearn.

In *Naked Truth: Your Journey from Nothing to Everything and Else*, Sandeep brings together a lifetime of inquiry into one distilled realization: **truth is not taught — it is remembered.**

Don't miss out!

Visit the website below and you can sign up to receive emails whenever Sandeep Chavan publishes a new book. There's no charge and no obligation.

https://books2read.com/r/B-A-EPGPC-DDBSG

BOOKS 2 READ

Connecting independent readers to independent writers.

Did you love *Naked Truth: Everything and Else*? Then you should read *Black Hole Don't Exist*[1] by Sandeep Chavan!

[2]

What If the Greatest Mystery in Physics Was a Misunderstanding of Stillness?

What if black holes are not collapsed stars devouring the cosmos — but collapsed understanding devouring science?

In this bold and paradigm-shifting book, **Sandeep J. Chavan**, independent researcher and originator of **Universal Energy Dynamics (UED)**, takes us on a revolutionary journey through the myth, mathematics, and misinterpretations that have shaped modern astrophysics.

Black Holes Don't Exist is not a conspiracy rant or science denial. It is a rigorous deconstruction of flawed assumptions — and a

1. https://books2read.com/u/boAMYA

2. https://books2read.com/u/boAMYA

courageous reconstruction of how the universe actually works. With piercing clarity, Chavan dismantles the narrative of singularities, gravitational sinks, and event horizons. In their place, he introduces a deeper truth: the universe is governed by **Structured Vacuum Energy (SVE)** and ripple dynamics — not force and fiction.

In this book, you will discover:

Why gravity isn't a pull, but a ripple gradientWhy "singularity" is not a physical event, but a failure of perceptionWhy motion is an illusion, and **stillness is the fundamental state of reality**How "black holes" are really **Ripple Saturation Zones,** not cosmic monstersWhy Hawking's paradox, gravitational waves, and collider data don't prove collapse — they prove **limits in coherence**How Einstein's **1939 paper** quietly warned against this entire detourAnd how the **observer's mind**, not the universe, creates the illusion of destruction

More than just a critique of mainstream physics, this is a blueprint for a **new scientific language** — one rooted in balance, not belief.

Whether you are a physicist disillusioned with paradoxes, a philosopher seeking structural truth, or a curious mind ready to shed scientific dogma — this book will change the way you see space, time, energy, and yourself.

There are no black holes. Only boundaries of understanding. It's time to step beyond them.

Read more at https://www.sandeepjchavan.com/.

Also by Sandeep Chavan

Chavan's UED research

CMGED Research Report

Chavan's Cosmic Continuum (C^3)

Frequency

Ripples of Reality

Reality Realigned

The Great Realignment

I – The Cosmic Leverage

Chavan's Cosmic Theory of Relativity (CTR)

Chavan's Theory of Light

Feeling Like a Physicist

TIME – The New Definition

$E \neq mc^2$: The Collapse of Classical Assumptions

The Chavan Codex

Chavan's Atomic Model

Chavan's Theory of Everything

Echoes of War: The Russia-Ukraine Conflict and Its Global Ripples

Shattered Lives

The War Economy

Borders Redrawn

Arms and Allies
Collateral Damage
Justice in the Ashes

Somewhere in the Darkness of the Cosmos
The Blueprint of the Universe
The Cosmic Breath
From the Edge of the Universe

The Eternal Sage Chronicles
The Sage and the Sovereign
The Family Pillar
The People's Flame
The Warrior's Soul
The Bridging Mentor
The Guiding Light

The Exodus
The Last Cycle: 2075

The Fundamental Theory of Life & Existence
Energy in Action: The Soul Demystified
The Active Mind
Energy in Connection
Death: The Disconnection
Subconsciousness: The Chosen Disconnection

Desires: The Engine of Life

The Restoration of Human Knowledge

Restoring Bhagavad Gita

The War of Thoughts

Cracks in the Known

The Human Game

The Divine Dilemma

The Original Intent

The Answers

Ultra Modern Life Philosophy

The Duality of Laughter and Tears

Take It Easy

Compare to Accept

Standalone

It's Not AI, It's AHI - Amplified Human Intelligence

The Decision Paradox

The Four Sapiens

Malicious Script of Indian Polity

The IIT Legacy & Global Impact

The IIT Legacy & Global Impact

You are just a Version of Your Original

Win the Game You Didn't Choose

Philosophy 2.0: A Structural Manifesto
Chaos: A Gentle Guide to the Unresolvable
Self-Help: New Syllabus
Rippling
Game: How to Play the Share Market
Out of Syllabus
Saffronized: The Code Behind the Crown
The Anger
Moksha: The Liberation
Chavan vs Einstein: The Universe Was Never Curved
Rocket Science: The Ripple Gambling
The Cosmos: The Still Universe
The Taste of Silence
Chavan's Post-Materialist Human Model
10-Second Rule
Chavan vs Hawking
It was Never a Wave Signal
Quantum Courtroom: The Trial of a Century
Chavan–Heisenberg Ripple Uncertainty Principle
The Great Disruption
Dark ≠ Hidden
Breath
Calculus of Life
The Earth Doesn't Move
Black Hole Don't Exist
Chavan vs Kant
Naked Truth: Everything and Else

Watch for more at https://www.sandeepjchavan.com/.

www.ingramcontent.com/pod-product-compliance
Lightning Source LLC
LaVergne TN
LVHW090315210926
848480LV00001B/255